The Potential Effect of Two New Biotechnologies on the World Dairy Industry

The Potential Effect of Two New Biotechnologies on the World Dairy Industry

Lovell S. Jarvis

Routledge
Taylor & Francis Group

LONDON AND NEW YORK

To Grandad and Ma

First published 1996 by Westview Press, Inc.

Published 2019 by Routledge
52 Vanderbilt Avenue, New York, NY 10017
2 Park Square, Milton Park, Abingdon, Oxon OX14 4RN

Routledge is an imprint of the Taylor & Francis Group, an informa business

Copyright © 1996 Taylor & Francis

Notice:
Product or corporate names may be trademarks or registered trademarks, and are used only for identification and explanation without intent to infringe.

A CIP catalog record for this book is available from the Library of Congress

ISBN 13: 978-0-367-29528-8 (hbk)
ISBN 13: 978-0-367-31074-5 (pbk)

Contents

Tables and Figures

Figures

Acknowledgments

I am grateful to John Mellor, who suggested the topic that led to this book, to the International Food Policy Research Institute (IFPRI), which funded the initial research, and to the American Breeders Service (IFPRI), which provided financial support to IFPRI to facilitate the initial study. I am grateful to Atanu Saha, who co-authored the projections analysis of milk production and consumption (Chapter 4) when he was a graduate student at the University of California, Davis, and to numerous other collegues who provided information, suggested additional readings, and/or read the draft manuscript and offered helpful comments, especially Gary Anderson, Jock Anderson, Michael Bishop, Robert Blake, Eric Bradford, L. J. "Bees" Butler, Daniel Chupin, E. P. Cunningham, Cornelis de Haan, Lucia Pearson de Vaccaro, Richard Perrin, Ed Rege, Hakan Sakul, Carlos Sere, Ken Shapiro, and Karl Zessin. I also thank Rene Hadjigeorgalis for research assistance, Susan Casement and Barbara Hagenbart for library assistance, Elizabeth Davis for editorial assistance and Claudette Oriol for preparing the manuscript for publication. None is responsible for remaining errors.

Lovell S. Jarvis

1

Introduction

Technological change is the main force leading historically to higher milk output, lower costs of production, and lower consumer prices. In recent years, the development of biotechnology has offered the potential for the emergence of new technologies that may further reduce the cost of milk production. This study analyzes the expected profitability, use, and effect of two new biotechnologies. One of these biotechnologies involves the administration to dairy cows of a genetically engineered hormone, recombinant bovine somatotropin (rbST), which can increase milk yields and feed efficiency. The other biotechnology involves the use of drugs to induce multiple ovulation in cows, the subsequent collection and fertilization of the eggs produced, and the transfer of the resulting embryos to other recipient cows. By permitting an increase in the intensity of genetic selection, multiple ovulation and embryo transfer (MOET) can also achieve a more rapid rate of increase in milk yields. The continual creation of first generation (F1) crossbred dairy calf embryos and their transfer to donor cows also offers an opportunity to increase milk production in some developing countries where crossbred dairy cows are the preferred choice of dairy farmers.

rbST and MOET are among the first two major biotechnologies created. That fact alone has gained them considerable attention. That they are used to produce milk, perhaps the most basic and one of the most widespread foods consumed by man, has heightened interest in their effects. rbST has generated particular attention because of concerns that its use might cause harm either to the cows who are injected with it or to the humans who drink the milk from these cows. Even if rbST is not harmful to human and animal health, there is also concern that its use might so increase milk supply that the disruptions created in the dairy market would bring greater social costs than benefits. MOET has been less discussed in the popular press, probably because MOET has not raised the same health issues. However, the long-term effects of MOET on milk production, milk price, and the structure of the dairy industry should be similar to those of rbST.

The two biotechnologies have potential to substantially affect the world's supply of milk during the next two decades. These biotechnologies are already being adopted in the United States. A number of other developed countries have begun adopting MOET, though most have postponed approval of rbST. What will be the eventual extent of the adoption of these biotechnologies in developed countries? Will these biotechnologies also prove profitable in developing countries and be adopted there too? These issues are explored here. The worst scenario from the viewpoint of developing countries would seem to be that the new biotechnologies will prove profitable only in developed countries. If so, although consumers in developing countries might benefit from lower international milk prices, dairy producers in developing countries would not be able to adopt them and thus would find their comparative advantage in milk production decreased.

On the other hand, high rates of population growth and the combination of high income elasticities of demand for milk and high rates of income growth should lead to substantially higher demand for dairy products in many developing countries. The development and adoption of rbST and MOET could achieve a substantial rightward shift in the supply curve of milk, permitting the growth of global milk supply to keep up with rising demand and perhaps even sustain a long term decline in consumer prices.

In addition, it appears that milk is a particularly important food from the viewpoint of human nutrition so that there should be special interest in technological developments that lower its cost. Research has shown that child nutrition in developing countries is positively associated with the amount of animal products in the child's diet (Allen et al., 1991). Although the way by which animal products affect child nutrition are not fully understood as yet, this finding suggests that the consumption of animal products conveys special benefits, probably via micronutrients, that are reflected in child development, perhaps cognitively and emotionally as well as physically. If this is so, increasing access to animal products is especially desirable. Food products of animal origin are usually more expensive than food products of vegetable origin, but milk is the animal product most likely to be accessible to most children in terms of availability and cost.

This study proceeds along three dimensions:

1. an analysis of the microeconomics of rbST and MOET in developed and developing countries, given those factors in each region that are expected to influence the profitability and adoption of these biotechnologies,
2. an estimate of these technologies' likely effect on milk costs, and
3. projections of possible supply/demand balances of dairy products in both regional and world markets under several different assump-

tions regarding the adoption of the two new biotechnologies in developed and developing countries.

Chapter 2 analyzes the profitability of rbST in developed and developing countries, and the factors on which such profitability depends. Chapter 3 discusses the profitability of MOET in developed and developing countries, including a specific analysis of MOET's use for the continuous development of first generation (F1) cross-bred calves for dairy use. Chapter 4 reviews the recent growth in milk production, consumption and international trade for major milk producing countries and for different regions, and projects future supply-demand balances with and without the increased production that the use of rbST and MOET might achieve. Chapter 5 contains conclusions.

Some advance knowledge of the book's conclusions may be helpful. It would be relatively easy to dismiss the current applicability of both rbST and MOET to developing countries because each requires a high level of management skill and of animal nutrition in order to achieve the increase in productivity that makes its use profitable. Few developing country milk producers can offer either of these inputs at the levels that will be required. In addition, developing countries already possess a large stock of conventional technologies that appear more appropriate to their existing resources than do these two biotechnologies, whose use taxes even the management skills available in developed countries. Investments in these conventional technologies or to facilitate their use will usually be more profitable in the foreseeable future than investments in the two new biotechnologies discussed here.

Nonetheless, it is crucial to understand the new biotechnologies, their effects, and the constraints on their use if appropriate policies are to be formulated in developed and in developing countries. From the viewpoint of this study, the issue is less one of safety and more one of profitability. Each of these two biotechnologies appears safe from a human health perspective, given the information currently available. Each poses some health risks for cows, but these appear relatively small and manageable.

It now appears that both rbST and MOET will be adopted and used on some farms both in developed and in some developing countries. Their rates of adoption will increase over time and it appears that their use will reduce the cost of milk production, mainly to the benefit of consumers. The technologies will be used more in developed countries than in developing countries for the foreseeable future. This fact does not mean that the two biotechnologies will soon become standard practice. Even in developed countries, most dairy producers will have to learn a considerable amount about the new technologies in order to make their use profitable, and they will have to significantly alter their management systems. In developing countries, more extensive management changes will be required, and the production context will also have to change. Most developing countries are beginning to make substantial progress in improving their production systems and in adopting more conventional technologies to bring down the cost of milk. This progress should be continued as it will provide greater benefits than efforts to use the new biotechnologies in the short run. Such progress is also the surest approach to the profitable use of rbST and MOET in the future.

2

Recombinant Bovine Somatotropin (rbST)

bST is a hormone that is naturally produced in cows' pituitary glands (Asimov and Krouze, 1937). bST is a homeorhetic control that coordinates the metabolism of many tissues in dairy cattle. It is scientifically known both as bovine somatotropin (bST) and also as bovine growth hormone (bGH). Among other effects, it triggers milk production in cows' mammary glands. Although bST was identified over a century ago, the high cost of its production restricted research and practical applications until recently.

Recombinant bovine somatotropin (rbST) is a genetically engineered synthetic analog of the natural hormone. Scientists identified the gene responsible for production of the natural bST and, using standard genetic-engineering techniques, duplicated the gene and spliced it into the DNA of E. coli bacteria. These bacteria can be multiplied in fermentation tanks where they secrete large amounts of the rbST (a protein closely similar or identical to bST), which then can be inexpensively extracted for injection into cows (Bauman et al., 1985; Roush, 1991).

The injection of rbST produces a biological reaction that is essentially the same as that which occurs in dairy cows that naturally produce high levels of bST. Indeed, some researchers have suggested that conventional breeding programs designed to increase milk yields in dairy cows may have been selecting for genes which lead to high levels of bST production. Under the appropriate conditions, cows injected with rbST produce more milk. Although they require an increase in nutrition in degree similar to that which naturally higher-yielding cows also require, rbST also increases the efficiency by which feed is converted into milk. rbST coordinates metabolic processes so that more nutrients are partitioned toward milk production during lactation and toward accumulation of lean tissue during growth (Bauman et al., 1989).

Four pharmaceutical firms applied to the Federal Drug Administration (FDA) for permission to market rbST. Approval of rbST was long pending because of controversy regarding its safety for humans and for cows. However, the Food and Drug Administration (FDA) of the United States determined that rbST is safe and approved the variety produced by Monsanto (Recombinant Methionyl Bovine Somatotropin — Sometribove) for on-farm use in November 1993. Sometribove became commercially available in the United States in February 1994. It is still uncertain how rapidly or to what degree on-farm adoption of rbST will occur, although several recent studies suggest that adoption in the United States will occur rapidly. The situation is different in Europe. For example, although the Committee for Veterinary Medicinal Products of the European Economic Community concluded that rbST is safe for use (Commission of the European Communities, 1993), the Economic Community (EC) nonetheless imposed a moratorium on the commercial use of rbST, primarily out of concern that its adoption would exacerbate EC milk surpluses.[2] Canada has also imposed a moratorium on the use of rbST.

The use of rbST has been opposed on the basis that its use will:

1. be dangerous to humans who consume the milk produced,
2. harm the cows to which it is administered, and
3. so dramatically increase milk supply that it will seriously decrease milk prices, causing special harm to poorer farmers who may not find it profitable to adopt rbST.

No evidence has been discovered that rbST is dangerous to humans, though debate on this issue continues (e.g., Rouse, 1991; McGuirk and Kaiser, 1991. The evidence regarding rbST's effect on the cows to which it is administered generally suggests no serious effects, (e.g., Burton et al., 1987; Bauman et al., 1989; Phipps et al., 1990, and Jordan et al., 1991), but again there is debate (e.g., Kronfeld, 1993). There is some indication that continued use of rbST is associated with an increase in mastitis, a worsening of some reproductive parameters like days open (Burton et al., 1990), and a reduction in the degree of improved feed conversion efficiency (McBride et al., 1990). These issues are among the primary concerns of rbST adopters in the United States as is discussed later.

Critics argue that rbST acts on lactating cells in the mammary gland through a messenger substance called insulin-like growth factor-1 (IGF-1), and that milk from rbST-treated animals contains higher-than-normal levels of IGF-1. The same substance acts as a messenger for human growth hormone in children and adults. Some research indicates that it may survive digestion and enter the human blood stream. If so, IGF-1 could cause significant health problems. However, other research indicates that IGF-1 is broken down in the digestive tract so that it could not harm humans. Similarly, although some claim that partially-digested rbST may be active in humans, FDA researchers have found that rbST itself is functionally indistinguishable from its natural counterpart and that the human body breaks down rbST in the digestive tract with no ill effects.

Many farmers have indicated reluctance to adopt rbST out of concern that their cows will be harmed and/or that milk consumption will drop. The voiced reluctance of farmers in some areas, like California, increased over time prior to the commercial release of rbST (Butler, 1995). Nonetheless, most studies have found that cows receiving rbST evidence no stress, lack of heat tolerance, health effects, nor any significant gap in calving interval (e.g., Bauman, 1987). Cows administered rbST may, however, evidence a higher level of udder infections. Because antibiotics are used to control udder infections, increased use of antibiotics to treat udder infections may cause some amount of the antibiotics to be passed to milk consumers if specified withdrawal times are not strictly observed. Additional use of antibiotics associated with rbST, it is agreed, could thus lead to higher human intake of antibiotics, reducing the effectiveness of the latter in fighting infection (Rouse, 1991). Other scientists argue, however, that milk is rigorously monitored for antibiotic residues and, as a result, there is an extremely low incidence of such residues in commercial milk (.04 percent, GAO Report, 1992).[3] These scientists maintain that use of rbST need not increase the presence of antibiotic residues in milk.

Consumers in many countries have expressed concern regarding the health effects of milk produced with rbST and some have indicated that they will reduce or stop milk consumption if milk is produced with rbST. Most farmers in the United States (e.g., L.J. Butler, personal communication) were concerned that this event could occur, particularly as they stood to lose whether or not they adopted rbST themselves. If milk supply were to increase and demand to decrease, the effect of rbST on milk markets could be extremely disruptive.[4]

A number of food companies, supermarket chains, and dairy co-ops have pledged not to use or sell milk that has been produced from cow herds in which rbST has been used (Rouse, 1991), and such companies, supported by consumer action groups, have called for the required labeling of milk which has been produced using rbST. The aim, of course, is to create two distinct products. Currently, however, there is no technology that can be used to determine whether rbST has been used to produce milk since it is impossible to distinguish between bST and rbST in milk. Thus, promises or assertions are the only means that brand-name products have to ensure consumers that their milk is rbST-free.

After approximately one year of commercial application, concerns over the effect of rbST on humans seems to be declining. It appears that consumers gradually will accept the use of rbST unless a significant health event occurs. In the absence of such disruption, however, the potential effects of rbST should depend primarily on the rate of adoption and the resulting increase in output, each of which depends on the price of milk. There is obvious potential for significant interaction between adoption, output, and price. Most previous assessments of the potential effect of rbST have focused on the United States. These analyses offer some insight into what can be expected in other countries, developed and less developed. Of course, the dairy sector in the United States has a significantly different structure from that prevailing in other countries, e.g., in both the EC and in India, so that any assessments of the U.S. dairy sector, while relevant, must be used carefully.

During the past 16 years there has been considerable discussion and some empirical analysis of the potential economic effects of rbST adoption (e.g., Fallert et al., 1987; Kalter et al., 1984; Marion and Wills, 1990; Magrath and Tauer, 1985; Butler and Carter, 1988; Boehlje, et al., 1987). The major issues of concern have been:

1. the expected profitability of rbST and, thus, the rate at which the new technology will be adopted by farms of varying characteristics,
2. the impact that adoption will have on aggregate milk output and on milk prices, and
3. the interaction between the rate of adoption and prices.

These discussions were largely speculative since there was no clear indication of how rbST would be priced, nor how its use might affect consumer demand for milk. Although adoption has recently begun, we still have only preliminary indications of how adoption, output, and demand will interact.

The Effect of rbST in Developed Countries

Estimates of how rbST will affect milk output require assumptions regarding:

1. the effect of rbST on milk production per cow,
2. the cost of rbST,
3. the amount and cost of other incremental inputs required for the effective use of rbST,
4. the milk prices that will prevail when decisions are taken regarding initial adoption of rbST,
5. the prices that will result after adoption has run its course, and
6. the minimum level of economic return required by farmers if they are to adopt rbST.

Yield Response to rbST

Yield trials for dairy cows administered rbST show considerable variability of milk production. For example, at the 1989 meetings of the American Dairy Science Association and the American Society of Animal

Science, reported production increases ranged from 2.5 percent to 30 percent, or about 1.3 lbs to 15.4 lbs of milk per day of treatment. These results have been confirmed in subsequent trials.

Much of the variation in response to rbST is generally attributed to differences in herd management, although genetic potential also appears to be a factor. It is clear that poorly-managed herds experience little or no increase in milk production from use of rbST, while well-managed herds show considerable response. The reported daily increase in one study for well-fed and well-managed cows ranged from 17 percent to 25 percent during the period rbST was being administered, which translates to a 10.5 percent to 15.5 percent increase in annual production (Marion and Wills, 1990). This is similar to the results obtained in other studies (e.g., Bauman, et al., 1989; Jordan, et al., 1991; and Phipps et al., 1990). Since the response is small during the early stages of lactation, it is not profitable to administer rbST until about the 90th day of lactation. Thus, rbST has a smaller effect on total output per lactation than on daily output when rbST is being administered.

It is still unclear the extent to which rbST will cause a constant proportional response in milk production from cows of differing genetic milk potential, and whether, for example, the absolute response will be greater for animals with higher base production. This issue is important because it will significantly influence which farmers will find it profitable to adopt rbST. Since, for example, there appears to be a positive correlation between farm size and cow milk yields in the United States, a positive response to rbST following genetic milk potential might inherently favor larger, wealthier farmers. Further, because the genetic production potential of dairy cows in developing countries is generally well below that of dairy cows in developed countries, rbST almost certainly will have a smaller absolute effect on milk yields in developing country herds.

Marion and Wills (1990) explored the implications of several different types of milk response to rbST use, which included:

1. a constant response (1,800 lbs/year—roughly 820 kg),
2. three different proportional responses (9 percent, 12 percent and 15 percent), and
3. a partly constant, partly proportional response.

A variant of their model is used in the analysis below.

Pricing rbST

The pricing of rbST depends more on marketing strategies and competition than on the actual cost of production. It is estimated that the marginal cost of rbST production is $0.05 or less, per unit administered

(Marion and Wills, 1990). Moreover, Marion and Wills indicated that there are large economies of scale in rbST production. Currently, only Monsanto has an approved product so there is little scope for intense price competition. It can be assumed that Monsanto will wish to charge a price well above the marginal cost of production to recover the amounts invested in research and development, product testing and approval, and marketing expenses, and provide for profits. The profits earned will be affected by the number of farmers adopting rbST, as well as the profit margin on each rbST dose. Given a positive variable cost of production and price-elastic, on-farm demand, manufacturers of rbST should prefer to sell their product at a relatively higher price to a limited number of producers who find it highly attractive, rather than lower the price to maximize adoption.

Marion and Wills utilized several basic assumptions to estimate the rbST price at which manufacturers will maximize profits. They estimated that the manufacturers' profit-maximizing price for sales of rbST to Wisconsin farmers would be between $0.25-$0.30 per daily cow dose of rbST, assuming that farmers require a fairly high rate of return to encourage adoption of rbST.[5] They noted that the profit-maximizing price is still higher if it is assumed that farmers are willing to accept a lower daily return.

Although hypothetical, Marion and Wills' estimates were illustrative and broadly consistent with the results of several other studies. Taken together, these studies suggested that the equilibrium market price of rbST in the United States was likely to remain between $0.25 and $0.40 per daily cow dose for some years. The higher the yield response to rbST, the more that response is proportional to the initial base of cow milk production, and the higher the manufacturers' variable cost of production, the higher is the manufacturers' optimal selling price.[6]

In fact, Monsanto initially priced rbST at $5.00 for a 14 day injection, or $0.36 per daily dose. In October 1994, Monsanto raised the price of rbST to $5.80, or $0.41 per daily dose, at the upper end of the expected range. In early 1995 Monsanto reduced the price to $5.00 again. Monsanto's decision to market rbST at the high end of the expected range, and even to try a price somewhat higher still, seems explicable by several factors. First, although the marginal cost of manufacturing rbST is low, Monsanto has accompanied the introduction of rbST with a very intensive marketing effort and is providing significant technical assistance to users.[7] Accordingly, the profit maximizing price is higher than predicted in Marion and Wills' study, which did not consider such costs.

Second, the heterogeneity of dairy producers (and their herds) throughout the United States is much greater than the heterogeneity found within Wisconsin, and it appears that Wisconsin farmers are relatively higher cost producers. This fact suggests that the profit maximizing price is somewhat higher than Marion and Wills estimated.[8]

Third, Monsanto may not have minded if a somewhat higher price for rbST resulted in a somewhat slower rate of adoption, given that slower adoption also implies a slower rate of increase in milk supply, and a slower decline in the associated price of milk. The potential for market disruption has been a major public concern, with potential for a political backlash, and Monsanto should have wished to avoid it.[9] However, Monsanto's decision to lower the price from $0.41 to $0.36 per daily cow dose in early 1995 indicates concern that the higher price was curtailing use of rbST sufficiently to reduce its own profits.

Cost Increases Associated with rbST Use

The use of rbST requires an increase in feed and other costs to support the rise in milk production. Marion and Wills noted that the relationship between incremental feed costs and incremental milk production achieved with rbST is similar to the general relationship found between feed costs and milk output. Using a relationship estimated in linear form for Wisconsin dairy herds (Leuning et al., 1987), Marion and Wills estimated annual feed costs per cow to be about $200 + $0.0289 per pound of milk produced annually for a "typical" hay and concentrate ration in Wisconsin.

Marion and Wills assumed that other variable costs would increase in response to the use of rbST by $0.87 per 100 lbs of incremental milk. These costs include the use of farm labor, power, veterinary services, and milk transport.[10]

The Profitability of rbST

Marion and Wills assumed that the price of milk would be $11.00 per hundredweight (about $0.24 per kg) when the adoption of rbST began, and that the price of milk would decline by 5 percent to 10 percent in response to an increase in aggregate milk output. Under these assumptions, Marion and Wills' best-guess calculations indicated that rbST would be economically feasible for use with cows whose pre-rbST yield was about 16,000 lbs of milk per year. These calculations are shown in Table 2.1.

Who Will Adopt rbST?

It has been widely expected that the use of rbST will have a differential effect on the profitability of different types of dairy farms within the United States (Fallert et al., 1987; Boehlje et al., 1987; Butler and Carter, 1988; Kaiser and Tauer, 1989). Considerable empirical evidence on the adoption and use of new technologies suggests a positive correlation between farm size and farmer education on the one hand, and the rate and extent of adoption of (profitable) new technologies on the other, and it seems likely that the same will hold for adoption of rbST. For example, Saha et al. (1994) find that, ex-ante, the willingness of Texas dairy producers to adopt rbST is positively related to herd size and education. Klotz et al. (1995) find the same of dairy farmers in California.

As there appears to be a correlation between farm size and milk yield, rbST might result in a greater expansion in milk production by wealthier farmers having larger farms with higher yielding cows. Smaller, poorer

TABLE 2.1   Incremental Revenues, Costs, and Profit Per Cow from Use of rbST in United States, Wisconsin, 1990

Incremental Revenue Assumptions:	
Milk price in $ per cwt.	$11.00
Annual production per cow without rbST	16,000 lb.
Percent annual response to rbST	12
Calculations:	
Incremental milk production:	
160.0 cwt. * .12 =	19.2 cwt.
Incremental revenue:	
19.2 cwt.* $11.00/cwt. =	$211.20
Incremental Cost Assumptions:	
rbST price per cow per day	.30
Increased returns to farm mgmt. per cow per day	.30
Milk hauling costs per cwt.	.20
Labor, power, breeding, and vet. costs per cwt.	.67
Feed cost increase per cwt.	2.89
Calculations:	
rbST price: $.30 * 215 days use per cow per year =	64.50
Returns to mgmt.: $.30 * 215 days =	64.50
Milk hauling: $.20 * 19.2 cwt. =	3.84
Other non-feed variable costs: $.67 * 19.2 cwt. =	12.86
Increased feed costs: $ 2.89 * 19.2 cwt. =	55.49
Total incremental costs per cow per year	$201.19
Net incremental profit per cow per year	$10.01
Percent reduction in unit cost of milk from use of rbST	0.5

Source: Based on Marion and Wills, 1990, and Perrin, 1991.

farmers could actually face lower incomes as higher aggregate milk output increases the downward pressure on milk prices. Concern that rbST would help wealthier owners of larger farms relative to poorer owners of smaller farms was one reason for opposition to the approval of rbST utilization. Adoption of rbST could also induce substantial regional shifts in production, related mainly to regional differences in farm characteristics (Boehlje et al., 1987).

Although several studies have concluded that the effect of rbST adoption is not likely to significantly disrupt milk markets, (e.g., Fallert et al., 1987, and Marion and Wills, 1990), relatively small changes in the assumptions that these studies have employed can produce different results having different implications. For example, although cows in Wisconsin average yields of about 14,000 lbs per lactation (similar to the U.S. average yield of about 15,000 lbs), many cow herds in the United States have average yields well above this level. California herds, for example, have average yields exceeding 19,000 lbs a year. Under the assumption that the milk response to rbST is proportional to the base level yields, owners of higher yielding herds should find rbST use more profitable. The range of expected increase in milk production is from 10 to 15 percent, and good herd managers may be able to achieve a milk output response to rbST of about 15 percent rather than the 12 percent Marion and Wills assume.[11] Similarly, the U.S. milk price in 1994, when rbST adoption began, was also higher than Marion and Wills assumed, at $12.50 per cwt versus $11.00 per cwt. Finally, while Wisconsin herds average about 50 cows, California herds average about 500 (Reed, 1994).[12]

If these alternative parameters are used as assumptions (maintaining the assumed price of rbST at $0.30 per daily cow dose), the Marion and Wills model indicates that rbST adoption would increase per-cow profits by nearly $200 per year (see Table 2.2), rather than the approximately $9 shown in Table 2.1. Similarly, while the potential annual benefit from adoption of rbST to an owner of a herd of 50 cows that could achieve such parameters would be $10,000, an average California herd would achieve an annual gain of $100,000.

This example points to the potentially great incentive for dairy farms having high initial milk yields, good management, and a large cow herd to adopt rbST. Although the example given uses optimistic assumptions, it emphasizes the role of good management, a high initial milk yield, and a large herd in determining rbST profitability. Under such conditions, use of rbST remains highly attractive even when rbST is priced at $0.36 or $0.41 per daily cow dose. Raising the price of rbST from $0.30 per cow dose, as assumed by Marion and Wills, to these levels reduces the net incremental profit per cow by about $13 and $24, respectively, but this is a rather trivial

TABLE 2.2   Incremental Revenues, Costs, and Profit Per Cow from Use
of rbST in United States, with Optimistic Assumptions

Incremental Revenue Assumptions:	
Milk price in $ per cwt.	$12.50
Annual production per cow without rbST	20,000 lb.
Percent annual response to rbST	15
Calculations:	
Incremental milk production:	
200 * .15 =	30.0 cwt.
Incremental revenue:	
30 * $12.50 =	$375.00
Incremental Cost Assumptions:	
bST price per cow per day	.30
Increased returns to farm mgmt. per cow per day	.30
Milk hauling costs per cwt.	.20
Labor, power, breeding, and vet. costs per cwt.	.67
Feed cost increase per cwt.	2.89
Calculations:	
rbST price: $.30 * 215 days use per cow =	64.50
Returns to mgmt.: included in net profit =	0.0
Milk hauling and other non-feed costs: $.87* 30 cwt. =	$26.10
Increased feed costs: $2.89 * 30 cwt. =	$86.70
Total incremental costs per cow per year	$177.30
Net incremental profit per cow per year	$197.70
Percent reduction in unit cost of milk from use of rbST	6.9

Source: Author's calculations, adapted from Marion and Wills, 1990, and Perrin,
1991. See text for clarification.

percent of the profits shown in Table 2.2, i.e., about 7 and 12 percent,
respectively.

Indeed, even if the price of milk is assumed to be $11.50 instead of $12.50,
and if rbST is priced at $0.36 instead of $0.30 per daily cow dose, incremental
cow profits from use of rbST are $155. It is hard to imagine that rbST will
not be quickly adopted by the owners of herds that have potential to earn
profits that are even close to these, with potentially significant effects on
regional as well as on national milk supply.

rbST Adoption and the Price of Milk

How much could the price of milk fall in response to rbST adoption? This is an issue of considerable debate. Perrin (1991) used the studies by Fallert et al., 1987, and Marion and Wills, 1990, to estimate the probable cost savings associated with rbST use. Perrin correctly pointed out that the experimental data available can be interpreted to suggest a rather small long-run shift in the industry supply curve as a result of the adoption of rbST. Assuming animal milk yields increase by 12 percent, costs also increase significantly with rbST use, so that the decline in the unit cost of production is fairly small. Perrin, using data reported in Fallert et al., 1987 and in Marion and Wills, 1990, calculated that the use of rbST will produce gross revenue gains of $199 and $211, respectively, compared with total cost increases of $118 and $201, respectively.[13] These pairs of changes in revenue and costs yield unit cost savings of only 4.4 percent and 0.5 percent. Perrin noted that such small unit cost savings suggest little potential for rbST adoption to cause large-scale disruption of the U.S. milk market. A study by Stennes, using a linear programming model, derived comparable unit cost savings of about 6 percent from rbST use in Canada.

Nonetheless, these studies probably underestimate the potential downward pressure of rbST adoption on market price. First, the cost figures in the Marion and Wills study are biased upward by their assumption that the costs of adoption include a required return to management. Such a return may be required by pioneering producers if they are to make the initial adoption decision, but once adoption begins and the technology becomes widely understood, this quasi-fixed cost should be competed away. Using Marion and Wills' figures, removal of the required return to management causes the cost savings from adoption of rbST to rise to 3.9 percent.

Second, the cost and production data selected by Perrin from the Fallert et al. and Marion and Wills studies relate to farms whose characteristics place them close to the break-even point for rbST adoption. Currently these may be the marginal farms from the viewpoint of milk price determination. However, they may not be the marginal farms after rbST is adopted. If adoption of rbST allows larger and more efficient farms to force smaller and less efficient farms out of production, industry savings could be significantly larger. For example, when Perrin's calculations are repeated assuming application of rbST to herds averaging 20,000 lbs of milk per year, a 15 percent production response, and a price of $12.50 per cwt, the reduction in unit costs achieved by using rbST rises to 6.9 percent.[14] See Table 2.2.

Third, the productivity gains which have been identified with rbST may increase after the technology has been in use. Research will seek to refine rbST use, particularly as regards animal and nutrition management, and there is likely to be a positive interaction between rbST use and that of other technologies, including animal breeding.[15] Assuming that rbST improves the industry feed-conversion rate, the cost of feed also may decline, further reducing the cost of milk. Most importantly, dairy managers may be encouraged to improve overall herd management more rapidly than would otherwise have occurred in order to be able to make profitable use of rbST.

The short-run price effect of rbST adoption could be even larger than the long-run effect if dairy sector adjustments occur rapidly. As a hypothetical example, assume that rbST is adopted by the largest 10 percent of dairy farms (and only such farms). Assuming that such farms, in the aggregate, produce 15 percent of U.S. milk output prior to the approval of rbST, and that such farms increase output by 15 percent as a result of rbST adoption, total U.S. milk output could rise quickly by 2.3 percent. Given a consumer demand price elasticity of about -0.3, the price of milk would have to decline by 7.5 percent to equilibrate the increase in output from this relatively small component of the U.S. dairy sector.

Use of rbST will become less profitable as the price of milk falls, but there is little likelihood that a declining price of milk will cut off use of rbST. Given the low cost of manufacturing rbST, Monsanto should reduce the price of rbST if needed to maintain significant rbST sales. Since rbST improves the efficiency with which cows convert feed to milk, so long as the price of rbST is sufficiently low and its use fairly simple, it would be profitable for well-managed herds achieving significant production increases to continue rbST use—even if the price of milk has declined significantly.

For example, using the Marion and Wills model, use of rbST still yields $15.70 in net incremental revenue per cow per year (versus no use) when the milk price has fallen to $5 per cwt, for herds averaging 20,000 lbs of milk per year, assuming a 15 percent milk output response to rbST, if rbST is priced at $0.10 per daily cow dose. See Table 2.3. This example is not intended to suggest that a sharp price decline will occur, but rather to indicate that market dynamics could allow for temporary undershooting of the long-run equilibrium milk price. As milk production begins to increase and prices to decline, rbST adoption could actually accelerate. rbST adopters would surely continue to use rbST and other dairy farmers should seek to adopt rbST as a means of survival.

Although rbST has potential to reduce U.S. milk prices by a substantial amount in the intermediate run, results through the first year of adoption suggest that adoption of rbST will not be sufficiently massive to disrupt markets in the short run. By mid-1994, about 20 percent of California dairy producers were using rbST (Butler, 1995). This is a significantly higher proportion of producers than had been predicted by pre-adoption surveys, suggesting that once rbST became available and appeared profitable, a significant number of producers decided to adopt. Nonetheless, users reported treating only a fraction of their herds (1 to 60 percent), averaging 25 percent (Butler, 1995). It appears that it has not been profitable to use rbST on the bulk of animals, even in herds with more favorable conditions.

Thus, although rbST users have larger than average herds, accounting jointly for about one third of California's dairy cows, only about 8.5 percent of cows were actually being treated. Assuming that rbST achieved a 12 percent increase in milk output for the cows treated, rbST would have achieved a 1 percent increase in California milk production during the first year.

TABLE 2.3 Incremental Revenue, Cost, and Profit Per Cow from rbST Use in United States Assuming Low Prices for Milk and rbST

Incremental Revenue Assumptions:	
Milk price in $ per cwt.	$5.00
Annual production per cow without rbST	20,000 lb.
Percent annual response to rbST	15
Calculations:	
Incremental milk production:	
200 cwt. * .15 =	30.0 cwt.
Incremental revenue:	
30 cwt. * $5.00 =	$150.00
Incremental Cost Assumptions:	
rbST price per cow per day	$0.10
Increased returns to farm mgmt. per cow per day	.30
Milk hauling costs per cwt.	.20
Labor, power, breeding, & vet. costs per cwt.	.67
Feed cost increase per cwt.	2.89
Calculations:	
rbST price: $.10 * 215 days use per cow=	$21.50
Returns to mgmt.: included in net profit=	0.0
Milk hauling and other non-feed variable costs: $.87 * 30 =	$26.10
Increased feed costs: 2.89 * 30 =	$86.70
Total incremental costs per cow per year	$134.30
Net incremental revenue per cow per year	$15.70
Percent reduction in unit cost of milk from use of rbST	1.4

Source: Author's calculations, adapted from Marion and Wills, 1990, and Perrin, 1991.

Further increases can be expected as the number of adopters increase (23 percent of producers indicate they are considering using rbST, though another 4 percent indicate that they tried it in 1994 and quit),[16] and as adopters apply it to a larger fraction of their herds. Still, it does not seem likely that the price of milk will fall more dramatically in the short run than in the longer run.

Fallert et al. and Marion and Wills each argued that adoption of rbST would not seriously disrupt the U.S. milk market. First, adoption appeared to be profitable only for farms that have somewhat higher-than-average production efficiency (as reflected in higher than average milk yields) and, second, adoption would achieve only moderate cost savings even for those firms which adopt. The calculations in this study suggest that use of rbST is likely to introduce larger cost savings for a significant number of U.S. dairy farms that are more efficient than those taken as being representative by Fallert et al., 1987, and Marion and Wills, 1990. Realization of these cost savings, if they occur, could allow such farms to expand production sufficiently. If so, milk prices in the United States could fall significantly in the intermediate run. Nonetheless, rbST is not a "magic bullet," destined to effortlessly allow an increase in output. Even for those farms having favorable characteristics, it appears that a relatively small proportion of cows can be treated profitably. This proportion should increase with time, but the increase seems likely to occur relatively slowly, thus avoiding any dramatic short run market disruption.

Adoption of rbST in other developed countries such as Canada, the EC, other Western European countries, and Japan depends principally on whether the use of rbST is permitted. If it is permitted, it seems difficult to imagine that it will not be used. However, dairies in these countries are generally smaller than those in the United States and, since the incentive to adopt is importantly related to the number of cows milked, a slower rate of adoption can be predicted in these countries. There is also greater concern for the impact of rbST on the profitability of smaller farms and also greater inherent resistance to the use of biotechnologies in these countries. These countries will be able to observe the experience of the United States with rbST and to implement additional regulations regarding rbST use if such regulations are deemed useful. To the extent that milk prices fall significantly in response to rbST use, there will be pressures in other countries to restrict the expansion of adopting farms, e.g., the use of farm quotas. Such regulations would reduce the incentives for adoption. Information obtained from the U.S. experience may also reduce risk and uncertainty, leading to more rapid approval of rbST in those countries than would otherwise occur. Barring emergence of a major health problem, one could expect to see rbST in use in essentially all developed countries within two decades, though the extent of adoption within such countries may vary greatly.

The Effect of rbST in Less Developed Countries

Although it appears that rbST will be rapidly adopted by many dairy farmers in the United States, and without much delay in other developed countries, significant adoption is not likely to occur in developing countries for a decade or two for lack of profitability. The profitability of rbST in developing countries will depend on the prices of rbST and other production inputs, the price of milk, and the incremental production response achieved.

The principal factors affecting the probable response of milk production to rbST in developing countries include:

1. the quality of farm management,
2. the availability of high-quality animal feed,
3. the climatic regime and the disease/parasite challenge,
4. the effectiveness of veterinary care,
5. the genetic potential of the existing herd, and
6. in the longer run, the opportunities for upgrading this genetic potential to respond more strongly to rbST.

Pricing rbST

The price of rbST in the developing countries is likely to be determined mainly by market conditions in developed countries. Developed countries account for two-thirds of world milk production. The production environment in developed countries is also much more favorable than that in developing countries. As a result, a high proportion of the rbST used during the next decade should be used in developed country markets, and the price in developed markets should strongly influence the price for rbST internationally. Only if the export of rbST from a low-priced market to a high-priced market can be controlled would the producers of rbST find it profitable to engage in price discrimination, selling rbST at a lower price in developing countries. The lower price is probably crucial to making rbST profitable in these markets. A lower rbST price should occur when the initial patents expire. This issue is explored later.

The Price of Milk

The reported "farm gate" price for milk in developing countries in the late 1980s ranged from about $0.12 to about $0.30 per kg (Seré and Rivas, 1987; Walshe et al., 1991), a range similar to that prevailing in developed countries during the same period. Although the international price of milk was sometimes as low as $0.12 per kg, domestic prices in developing countries usually exceeded the international price by a significant margin due to domestic protection. In the analysis that follows below, a price of

$0.24 per kg is used. Clearly, a higher or lower price would change the incentives for adoption.

Potential Yield Response to rbST

Most developing countries are located in tropical regions having a relatively unfavorable milk production environment including high temperature and humidity, low quality pastures, and a high incidence of disease and parasites (McDowell, 1985; Smith, 1985). The cattle populations in these regions are composed predominantly of Zebu (*Bos indicus*) breeds and their crosses. Such cattle display considerable tolerance of the climatic, nutritional, and disease challenges present in these areas. However, their genetic capacity for milk production is limited even under more favorable conditions.

Not surprisingly, milk yields in less developed countries are generally significantly lower than those in the United States, Canada, and Western Europe. The magnitude of the difference is quite large, as shown in Table 2.4. The average annual yields in Africa and the Middle East are about 500 kg, only one tenth those achieved in the United States and Canada. Yields in Latin America and Asia are about 1,000 kg and 2,500 kg, respectively. Although milk yields vary considerably across regions and even across herds within the same region (or country), the average yields reported are representative of the bulk of herds in these areas. Insofar as herd genetic mix, management practices, and other inputs are largely determined by environmental conditions, the average yield data provide a useful indication of overall production conditions in each region.

Few data are available regarding the production response which can be achieved in developing countries through the use of rbST. This response can be expected to vary widely because of the variation in the type of cattle, feed, and quality of management locally available. However, if the response to rbST were approximately proportional to the pre-rbST yields (which average only about one tenth to one half those achieved in developed countries), the absolute increment achieved by rbST would be much smaller than is achieved in the developed countries. Accordingly, the use of rbST would be less profitable. Indeed, given that rbST appears to be profitable in the United States only for dairy herds having cow milk yields of approximately 6,000 kg, it immediately seems unlikely that use of rbST will be profitable for most herds in developing countries.

High quality management, improved nutrition and timely veterinary care also have been shown to be crucial to profitable application of rbST in developed countries. These inputs will be at least as important in developing countries where the environmental challenge is greater, yet these crucial inputs are much less frequently available in developing countries.

TABLE 2.4   Cow Milk Yields, by Region and Period (kgs.)

	1961-1970	*1971-1979*	*1980-1988*	*Annual Avg. Yield Growth*
DCs	2,628	3,004	3,365	1.4
LDCs	653	702	753	0.8
U.S. & Canada	3,704	4,675	5,715	2.4
Japan	3,948	4,145	4,891	1.2
EC	3,157	3,561	4,118	1.4
Western Europe	2,968	3,543	4,084	1.7
Australia & New Zealand	2,496	2,891	3,291	1.5
Eastern Europe	2,029	2,358	2,652	1.5
South and Southeast Asia	1,003	1,579	2,497	5.0
Africa & Middle East	516	531	552	0.4
Latin America	1,028	1,045	1,031	0.0

Source: Author's calculations, based on FAO Agricultural Production Tapes.

Management and Nutrition as Limiting Factors

Little is known about how rbST will interact with genetic potential in terms of resistance to ambient pressure caused by factors such as temperature, humidity, disease and parasites. Use of rbST has achieved significant milk increments in numerous Arizona dairy herds, despite the high ambient temperatures prevailing in Arizona. However, Arizona dairy herds generally have some of the highest yields in the United States, reflecting outstanding genetics and management, and Arizona is an area of generally low humidity. Large output increases also have been reported in response to the use of rbST in Mexico, Brazil, and Zimbabwe during recent years, but mostly in regions where ambient temperatures are lower than those in Arizona. In hotter climates, the response to rbST depends on cow comfort as well as feeding practices, with higher response occurring where cows are cooled with sprinklers and fans. Thus, rbST can be used in areas of high temperature if nutrition, genetics, infrastructure, and herd management are appropriate.

Preliminary evidence also indicates that it is physiologically possible to obtain higher proportional incremental outputs from use of rbST from initially low-yielding *Bos indicus* cows than from *Bos taurus* cows, provided that nutrition and herd management are good (Phipps et al., 1991). These two requirements, however, will preclude profitable adoption for most developing country herds.

Research in the developed countries points to the importance of high-quality nutrition for optimal response to rbST (e.g., Chalupa and Galligan, 1988; Burton et al., 1990). For economical use of rbST, cattle must be provided with a balanced ration that they can and are inclined to ingest, tailored to fit the expected increase in milk output. If too little feed or an improper ration is given, output will be limited; if too much feed is given, feed will be wasted, increasing costs. Management must be capable of making the required adjustments in nutrition and otherwise maintaining the herd in good condition.

Limitations on nutrition and management constrain the milk yields achieved from the livestock in developing countries, many of which have the genetic potential to produce significantly more milk than they do (McDowell, 1985; Vaccaro, 1994). The use of rbST itself will not improve either nutrition or management in the developing countries, and it will be impossible to improve herd nutrition and management in most areas in the manner required for profitable use of rbST.

Few dairies in developing countries have record keeping systems which will permit an accurate monitoring of feed inputs and milk outputs, and which is needed if appropriate decisions are to be made regarding which cattle to be injected with rbST, and how much to feed them. Equally important, high-quality feed—whether concentrates or pastures—is often not available or is prohibitively expensive in many developing country regions. In the United States, feed rations are largely based on concentrates because these have a high nutrient density and can be more easily fed and consumed. The nutrient density of feed ration is important since cows' appetites often limit their intake below the desired level, particularly in hotter weather. However, dairy cows in most developing countries are fed mainly on pastures. Research indicates that properly managed herds whose nutrition is from grazing can achieve a profitable response to the administration of rbST similar to those achieved by herds that are fed concentrates, but it is more difficult to obtain an adequate response.[17]

Potential Use of rbST on Bos Taurus Cattle

Bos taurus dairy cattle in developed countries have milk yields ranging from 2,000 to 10,000 kg per year (4,400 lbs to 22,000 lbs), while *Bos indicus* dairy cattle in developing countries have milk yields ranging from about 500 to 2,500 kg per year. Substitution of *Bos taurus* for *Bos indicus* cattle has been perceived as a means of increasing milk yields in developing countries. Many purebred dairy cattle were imported into the developing countries after World War II (Vaccaro, 1979; Jasiorowski et al., 1988). However, the efforts to introduce purebred *Bos taurus* animals into most developing country regions proved disastrous (Brumby, 1979; Vaccaro, 1994). *Bos taurus* dairy cattle perform poorly in most developing country regions because there is an inherent incompatibility between these animals and the developing country context into which they are placed. Animals with the genetic potential to produce high milk yields are genetically unsuited to the harsh environment, disease/parasite challenge, and poor nutrition available in developing countries (Vercoe and Frisch, 1980). Purebred *Bos taurus* cattle suffer high morbidity and mortality, have longer calving intervals, require more feed (and of higher cost), and produce less milk than do local varieties.

Purebred *Bos taurus* breed cows have been introduced successfully in some, usually highland, regions where the environmental context is more similar to that faced in the developed countries. However, these herds almost certainly are accounting for a steadily declining share of milk production in the developing countries because they have proven uncompetitive with milk produced from local and crossbred animals grazed on lowland pastures (Vaccaro, 1994; Seré and Rivas, 1991; Nicholson, 1990; Jarvis, 1986). Most milk in developing countries is produced by *Bos indicus* purebred and crossbred cattle types. In several Latin America tropical regions, a substantial number of beef cattle are milked. Although cow yields are low, total milk production has been rising rapidly. Purebred *Bos taurus* cattle in the more favorable regions can produce higher yields, yet the land in such regions can produce much higher output if dedicated to more valuable crops. The use of hardy breeds, often *Bos indicus* breeds and crossbreeds, can produce milk in lowland areas which, at least at present, have lower opportunity cost.

Potential Use of rbST on Bos Indicus and Crossbred Cattle

Bos indicus cattle used for dairying have yields ranging from about 150 kg to 2,500 kg per year, although, more commonly, yields range from 500 kg to 1,500 kg. Genetic selection has the potential to increase the milk yields from such cattle (Vacarro, 1994), but crossbreeding *Bos indicus* with *Bos taurus* cattle has been a means of raising milk yields more rapidly, while still retaining substantial hardiness. For example, a cross between a *Bos indicus* cow with a potential yield of 1,000 kg and a *Bos taurus* bull (inexpensive semen is available in the developed countries with a potential yield of 8,000 to 9,000 kg) should yield an average of the two plus the effects of heterosis, which are expected to be positive (e.g., Cunningham, 1987). Heterosis is discussed in some detail in Chapter 3.

Crossbred animals perform well in most tropical developing countries. Their superior yields are a major explanation for the rapid expansion in the use of crossbred animals in many developing countries (Cunningham and Syrstad, 1987; Turton, 1981 and 1985; Vaccaro, 1994; Jarvis, 1986; Holmann et al., 1990a). Still, that most crossbred cows achieve yields well below their genetic potential is further evidence of the external limitations. For example, while first generation (F1) crossbred animals theoretically could yield at least 5,000 kg, their yields commonly reach only about half this amount (Cunningham and Syrstad, 1987; Turton, 1981 and 1985). Poor management, nutrition, and the environment challenge limit milk yields to levels well below genetic potential. These same problems are expected to limit the use of rbST.

Milk Production in the Latin American Tropics

Vaccaro (1994) emphasized the importance of fitting the production system to local resources, a tendency which led Latin American producers to emphasize intermediate as opposed to highly intensive systems (Preston, 1977; Seré, 1983) based on grazing or crop residues with strategic supplementation (Preston and Leng, 1987) using dual-purpose or multi-purpose cattle of local or European x local origin (Mukherjee, 1990; Jarvis, 1982 and 1986; Turton, 1981 and 1985; Cunningham and Syrstad, 1987). Systems relying on low to intermediate levels of inputs make the best use of locally available feed and animal resources in most developing regions, and also fit with limitations of capital, management skills, and support infrastructure (Preston, 1977).

The comparative advantage of the Latin American tropics in ruminant livestock production has been based mainly on the availability of large amounts of relatively low quality, but inexpensive pasture. Latin America's lowlands are characterized by high temperatures and seasonally high moisture. Tropical grasses produce large volumes of forage of low to intermediate quality (Seré and Rivas, 1991). Such pastures have low digestibility and frequently exhibit nutrient imbalance (Minson, 1981; Preston and Leng, 1987), resulting in low food consumption by the animals under conditions of heat stress (Vaccaro, 1979).

Despite such an apparent environmental disadvantage, the dairy industry in these areas has been growing fairly rapidly in recent years (see Chapter 4). Milk can be produced more economically by using local breeds and crossbred cows to graze these tropical lowlands, as compared to the use of purebred *Bos taurus* or higher grade crossbred cows on improved pastures in many highland tropical areas where the cost of land is much higher. This lowland production system achieves relatively low per animal yields of both milk and beef per cow, but with fewer inputs and lower cost per unit output, as compared to the highland systems that achieve higher per animal milk and meat yields.

Table 2.5 (Vaccaro, 1994) provides data on the comparative costs of production and on net profit from several different production systems in Latin America. The data show that:

1. the use of 50 percent European-zebu or European-local cross cows is more profitable than the use of purebred or higher grade European cows, even when management is of high quality,
2. a system with an intermediate level of inputs and lower quality management is consistently more economical than a high output system with higher quality management,

TABLE 2.5 Relative Costs and Profits from Crossbred Cattle of Intermediate Levels of European Breed Inheritance, Compared with Other Genotypes in Different Production Systems in Latin America (50% European or Local Cross = 100)

Country/System	Breed group	*Relative*	
		Costs	Profits
BOLIVIA			
Pasture + Supplement	Local cross	100	100
	European	190	–300
Confined	European	210	–300
BRAZIL			
Low level management	25% European-zebu		38
	50% European-zebu		100
	75% European-zebu		53
	European		–30
High level management	25% European-zebu		–27
	50% European-zebu		40
	75% European-zebu		38
	European		30
VENEZUELA			
Lowland, medium level management	50% European-zebu	100	100
	75% European-zebu	109	83
Highland, high level management	European	167	70

Source: Vaccaro, 1994, based on Wilkins *et al.*, 1979; Madelena *et al.*, 1990b; and Holmann, 1990.

3. lowland production systems are more profitable than highland
    systems.

Table 2.7 (Seré and Rivas, 1987) shows the total number of dairy cattle
in most countries in the Latin American tropics, the percentage of those
which are classified as dual-purpose, and the proportion of total milk
supplied by the latter. Dual purpose cattle supply more than half of the
milk in Brazil, Guatemala, Honduras, Mexico, Nicaragua, and Colombia,
with appreciable proportions in most other countries. Latin America has
manifested no increase in average cow milk yields in recent decades
despite sustained growth in total output. This is consistent with a steady
shift of milk production toward the low-input system which has been
documented in Brazil, Colombia, Venezuela, Costa Rica and Mexico, among
other countries (Preston, 1976 and 1977; Jarvis, 1986; Seré et al., 1990;

TABLE 2.6 Importance of Dual Purpose Milk Production in the Latin
American Tropics, 1984

Country	Dairy Cows (thousand)	Yield per Cow (kg/year)	Dual Purpose Cows as Percentage of Dairy Cows*	Milk of Dual-Purpose Cows as Percentage of Total*
Bolivia	56	1418	54	19
Brazil	14700	714	89	63
Colombia	2800	1000	75	38
Costa Rica	270	1259	62	25
Dominican Republic	229	2009	25	6
Ecuador	720	1375	56	20
El Salvador	261	954	77	41
Guatemala	400	825	84	51
Honduras	430	651	92	71
Mexico	8900	812	84	52
Nicaragua	200	625	94	75
Panama	96	958	77	40
Paraguay	87	1897	30	8
Peru	675	1156	67	29
Venezuela	1387	1072	82	57

* Calculated by imputing an annual production of 2500 kg per specialized dairy
cow and 500 kg per dual purpose cow (750 kg per dual purpose cow for Venezuela).

Source: Seré and Rivas, 1987.

Nicholson, 1990; Frankel, 1982). Increasing milk production in the Latin American lowland tropics has been highly dependent on public investment in infrastructure development—roads, main trunks as well as rural access roads, collection and cooling facilities, dairy processing facilities, and distribution facilities (Seré and Rivas, 1987).

The majority of cows in tropical Latin America are hand milked with the calf present. The calves nurse initially to stimulate milk let-down and are allowed one quarter and/or the residual milk. This system is profitable because labor (and pasture) is cheap. Labor is relatively poorly skilled, however. The system would not lend itself to the needs of a more complex production system where greater management precision is required, as would be the case if rbST were used.

Milk production in these lowland, intermediate-level input systems is usually highly seasonal, in keeping with seasonal fluctuations in rain and available forage. Output is constrained in some areas by a pronounced dry period. Producers graze hardy breeds having low yields on pastures of relatively poor quality. Farm-gate milk prices are relatively low. The average prices received, which are heavily weighted by the prices occurring during the peak season of production, are usually lower than the average prices received in more favorable areas where production is more constant throughout the year. Milk collection and marketing systems are poorly developed in some areas. All of these factors combine to reduce the profitability of technologies like rbST. As a result, the use of rbST is not likely to be profitable for the vast majority of producers in the Latin American tropics in the foreseeable future.

Milk Production in Asia

Dairying in Asia is mixed, relying on two broad types of production systems (Singh, 1987). Many cows are kept by small-scale farmers as multiple-purpose animals that are used to produce milk as well as other outputs such as manure, draft power, and calves. These cows are fed principally on inexpensive, low quality feed that can be obtained from crop residues, common pastures, and roadside grazing, or from cut pasture and forage, with strategic supplementation where this is justified by the price of milk. The profitability of these cows is based on the farmers' ability to produce a variety of outputs in limited amount, each of importance to the farm system (Jarvis, 1982). Such animals convert crop residues and local pastures, which would otherwise have little or no value, into a variety of usable products. It is often economical for these farmers to produce a small amount of milk from these cows, even though yields are low. The cow is able to produce a relatively small amount of milk along with several other outputs. Each output provides value to the producer. All outputs are needed to make the use of the animal profitable. However, it will be difficult or impossible to alter this system in a way that would permit the profitable use of rbST to produce significantly higher levels of milk.

The animal breeds used in these systems are hardy—capable of surviving on limited nutrition in the face of a fairly severe disease and parasite challenge, in hot and sometimes humid climate, and where management inputs are usually low. A high milk output response to rbST is possible only if the farmer can provide a constant, higher level of management, nutrition, and appropriate veterinary care (given the persistent disease and parasite challenge faced). Further, given the interactions between uses of such animals, efforts to obtain higher milk outputs will probably be unsuccessful unless a move toward specialization in the production of milk is possible. And the farming system may not permit this. A different feed regime also would be needed, requiring either that the farmer produce forage crops or purchase concentrate rations.[18] There is currently no research that demonstrates that multi-purpose cattle are unable to respond adequately to application of rbST, but it seems highly improbable.

Other farms in Asia specialize in dairy production. Such farms depend increasingly on crossbred cattle and are fed mainly concentrates. These specialized cows have significantly higher milk yields than those of the multiple purpose cattle described above. Farmers often have one or two such cows, and herds rarely are large. Such cattle still account for a relatively small proportion of total dairy herds. Slade (1987) notes that the adoption of crossbred animals in India has been slow, precisely because their merits have not been clear to many farmers. Crossbreeds are thought to be poorer draft animals, requiring more fodder and being able to carry out less work. Small-scale farmers often lack fodder to feed them adequately. Crossbred cows may also suffer higher mortality and lower calving rates due to insufficient hardiness in what remains a highly stressful environment. The climatic, disease, parasite threat is still significant in most of the region.

Infrastructure, both physical and social, are likely to be more adequate in areas where milk production is important. In many areas in India, for example, roads are more adequate. Farm density and strong government support has permitted organization of cooperatives able to provide technical assistance and at least the basic inputs, as well as to collect, process, and distribute milk with greater certainty to the farmer. In most of Asia, farmers can increase output significantly simply by improving nutrition, management, and other limiting inputs. Nonetheless, the use of rbST would overtax their management capabilities and the resources available, even in most areas where crossbreeds are maintained. Although management of the existing farm system is often good, farmers have limited education and would not be able to handle a complex system like rbST in the foreseeable future.

Milk Production in Africa

In Africa, pronounced seasonal rains lead to seasonal pasture production in many areas where dairying is undertaken. Dry season feed is a major constraint on milk output, although use of rbST would require provision of an improved feed ration throughout this period. Africa has a shortage of grains for human consumption, and few farmers have sufficient labor or capital to cultivate grain or forage for animal feed. Current predictions indicate that Africa is unlikely to produce adequate energy and protein for livestock feed in the foreseeable future (Winrock International, 1992). Local price ratios generally do not permit profitable supplementation to achieve higher milk yields. Farmers are often distant from commercial markets. Milk marketing and processing networks are underdeveloped and often unreliable in terms of offering a secure outlet for production (Shapiro et al., 1991). Markets are segmented by unreliable and expensive transport. Production is highly seasonal, and additional output of the sort which would be stimulated by rbST during the period of high production is often of much less value. Processing plants and cooling facilities at collection centers are vulnerable to problems with electrical supply, machinery breakdown, limited spare parts, and insufficient packaging materials (Walshe et al., 1991; Winrock International, 1992).

Parasites and viral animal disease are a particularly serious challenge in Africa, increasing mortality and morbidity, lowering milk yields and increasing farm risk. Many are vector transmitted, widely distributed, and not controllable with effective or easily administered vaccines and chemotherapeutic agents (Winrock International, 1992). Veterinary personnel are limited in Africa and the supply network for inputs is poorly developed (Frankel, 1982; de Haan and Nissen, 1985; Winrock International, 1992).

Although farmers are often skilled in the use of existing technology—which is also adapted to the difficult conditions faced—they are not suited by training or experience to handle more intensive production systems. For example, considerable milk in Africa is produced by pastoralists who are known to be among the world's most skilled herders (e.g., Dahl and Hjort, 1976; Jarvis, 1991). However, these pastoralists are skilled at producing milk within a harsh production context, often by moving continually with their animals through semi-arid regions in search of pasture. Through diligent effort, they eke out small milk yields from their cows, sufficient for their own consumption and, sometimes, a small surplus for sale. However, even where herdsmen have begun to settle into sedentary farming systems, they have not become accustomed to nor skilled in the use of technologies which utilize high levels of inputs to produce high levels of milk.

In Ethiopia, a country in which specialized milk production has a long history, most milk is consumed within rural areas, mainly by the house-holds of producing families (Mbogoh, 1991). Milk sold, processed, and distributed through an organized marketing system accounts for only about 2 percent of total production. The road network in Ethiopia is poor in the rural areas, and over 80 percent of Ethiopia's peasant producers are at least 5 km from the nearest all-weather roads. Thus, although Ethiopia produces a considerable amount of milk, the use of rbST to increase milk production for commercial sale is currently inconceivable, except on a mi-nuscule and experimental scale.

In significant parts of Africa the threat of trypanosomiasis[19] makes the maintenance of cattle impossible or limits their level of output to low levels. In most areas where livestock are maintained, the low level of animal nutrition, the parasite-disease challenge, and the hot, humid climate have led to the adoption of low-input management systems which are wholly incompatible with the requirements of the high-input system required to utilize rbST.

The number of problems plaguing milk production in Africa are large and so serious that it is very unlikely that rbST would find significant use during the next 20 years. South Africa is a clear exception to the situation portrayed, having a much better developed dairy industry. Parts of Kenya, Zimbabwe, and areas in several other countries are also possible exceptions, though atypical. For example, the smallholder dairy development scheme in Kenya has been successful primarily because of massive support received from the government. Such support involved the development of an effective artificial insemination scheme for upgrading *Bos indicus* cattle, government intervention in milk pricing and marketing, and rural development activities that enhanced the market for farmers' milk production (Mbogoh, 1991). Also important was the development of a network of good rural access roads that feed into the main all-weather highways, the provision of credit for production investments, the establishment of dairy processing facilities in rural areas, and the development of functioning artificial insemination, veterinary and livestock extension services. Where governments are willing to make such investments, often for social as well as economic reasons, investments in conventional dairy technologies are likely to provide higher returns than efforts to use rbST.

The Importance of Management Skills

Dairy farms throughout developing countries are usually small farms in which a poorly educated owner and his family care for a few animals. In Latin America there are a number of larger farms in which a better educated owner hires relatively unskilled workmen to husband his animals, with the owner supervising the work or employing an administrator. Thus, whether on small farms or large, the direct care of dairy animals in each of these farming systems is entrusted to individuals who are skilled in low-input animal management, but who have neither the experience nor the education that would allow them to efficiently implement a high-input animal management system of the type required to make adoption of rbST profitable.

In most developing country regions, dairy productivity is reduced by a severe disease-parasite challenge. This challenge can be controlled through improved veterinary care, including vaccinations and use of dips, but improvements to the basic environment, such as higher quality water and external control of pests, are also highly important if less hardy animals capable of responding to rbST are to be used. Veterinary delivery can be improved and its costs reduced. Animal husbandry and dairy management can be improved through formal training of owners, managers, and workers and also through informal education, including technical assistance. Technical assistance can be provided by public institutions or, frequently with excellent results, through cooperatives. Farmer education is usually related to farmers' ability to successfully adopt new technologies, and to total returns.

Conditions Under Which rbST Use May Be Possible

It is unlikely that rbST can be profitably administered to the vast majority of developing country milk cows, but the use of rbST may be profitable in exceptional circumstances. Several examples are given below. Examples like these could occur more frequently in the future as nutrition and management improve and as the price of rbST declines. Once Monsanto's patent expires, the price of rbST should fall dramatically, to about $0.05 per daily cow dose.

Few rbST trials have been undertaken in less developed countries, but the studies published suggest that rbST significantly increases yields in buffalo (Ludri et al., 1989; Ferrara et al., 1989) and in *Bos indicus* and *Bos indicus-Bos taurus* crossbred cattle (Phipps et al., 1991). These are the dairy animals most commonly encountered in less developed countries. *Bos taurus* cattle in less developed countries have also shown a significant response to rbST when the production conditions have been adequate.

For example, Phipps et al. (1991) report on the results of applying 500 mg of rbST (Monsanto's prolonged release formulation, Sometribove) to three types of dairy cattle in Zimbabwe, including 24 dairy crossbreeds (Mashona/Nkone x Friesian) and 115 *Bos taurus* cows (48 Holstein, 42 Friesian and 25 Jersey).[20] Application of rbST to the *Bos indicus* and crossbred cattle was carried out at the Henderson Research Station, while application to the *Bos taurus* cattle was carried out on three private farms. The pre-treatment milk yields averaged approximately 150 kg, 2,500 kg and 7,400 kg for the *Bos indicus* cattle, crossbreeds, and Holstein cattle, respectively.[21] In each trial, management and nutritional inputs appear to have been high.

The *Bos indicus* cattle were divided into two groups based on the stage of lactation. Cows in group 1 had previously completed 65-95 days of lactation and cows in group 2 had completed 96-125 days. As noted previously, rbST has not significantly increased milk yields for *Bos taurus* cows during the early stages of lactation and rbST is generally administered only after 90 days of lactation. However, *Bos indicus* have a much shorter lactation. For this reason, rbST was provided slightly earlier in the lactation to the *Bos indicus* cattle.[22]

The mean milk yield for both control and treated cows in group 1 prior to treatment was 1.27 kg/day, and in group 2 it was 0.70 and 0.73 kg/day, respectively. All treated cows received seven subcutaneous injections of Sometribove (500 mg) at 14 day intervals. All cows grazed nitrogen-fertilized Bermuda grass pasture, without concentrate supplement. Use of rbST led to a 278 percent increase in milk yield over a 98 day treatment period, from 0.45 kg/day (1 lb) for the controls to 1.72 kg/day (23.78 lb) for the treated cows, an increase of 1.27 kg/day (2.78 lb). The authors noted that treatment appeared also to have increased lactation length, though this was not formally analyzed in the study.

Despite the high percentage increase in milk yield during the treatment period, per cow milk production per lactation rose much less. Daily milk yields were falling rapidly when rbST began to be administered. Although use of rbST achieved a high percentage increase, the absolute increment was small, and was achieved for a relatively short period. The use of rbST increased total milk production by approximately 35 percent, relative to that of the controls. This increase is only about 13 percent as great as the daily increment achieved when rbST was used.

Using a variant of the Marion and Wills (1990) model, the information provided in Phipps et al. (1991) can be used to calculate the expected profitability of applying rbST to these *Bos indicus* cattle. Several assumptions are needed. First, the local milk price is assumed to be $0.24 per kg ($11.00 per cwt), about the average reported "farm gate" milk price in developing countries, which in the 1980s, ranged from about $0.12 per kg to about $0.30 per kg (Seré and Rivas, 1991; Walshe et al., 1991). This price is similar to that prevailing in the United States during the same period.

Second, the biological relationship between incremental milk output and incremental nutritional (feed) needs are assumed to be the same as those found in Wisconsin, but the cost of feed is assumed to be only two-thirds as high. Pasture on a per area basis is generally much cheaper in developing countries than in developed countries, but it is also usually of inferior quality

(Van Soest, 1982) so that the cost per unit of nutrition is only somewhat less costly (Nicholson, 1990) in developing countries. Other assumptions regarding the cost and quality of feed are equally plausible insofar as these vary widely across developing countries. It would be impossible to model this relationship for all types of animals in all types of developing country locales and the attempt here is simply to obtain an indication of the magnitudes involved.

Third, the increase in "other" costs associated with an increase in milk production (e.g., labor, veterinary care, electricity, and milk hauling) is assumed to be three-fourths as high per unit of milk produced as was assumed in the Marion and Wills study. Wages of agricultural workers are low in developing countries. For example, agricultural wages throughout Latin America average about one tenth to one twentieth those in the United States (in US dollar prices at market exchange rates). However, laborers are considerably less well trained and less productive, so that unit labor costs range from about one fifth to about one third as much as in developed countries. Further, the cost of transportation, electric power, and veterinary supplies are often similar to the costs faced in developed countries, although again there is substantial variation among and within developing countries.

Fourth, rbST is assumed to be priced at $0.36 per daily cow dose, the same as in the United States. If manufacturers of rbST are able to price discriminate, they should find it profitable to sell rbST at a lower price in developing countries. The effect on profitability of lowering the price of rbST is considered later.

With these assumptions, the profitability of administering rbST to the Mashona and Nkone *Bos indicus* cattle in Zimbabwe can be calculated. Because data is available regarding the average increment in milk production during the period when rbST was administered, the per day profitability of rbST use is calculated. Administration of rbST increased output by 2.78 lb per day. See Table 2.7. Valuing this milk at $11.00 per cwt., or $0.11 per lb, the daily benefits were $0.31 per animal. Assuming a cost for rbST of $0.36 per daily cow dose, costs of hauling milk of $0.15 per cwt, costs of labor, power, and veterinary costs of $0.50 per cwt, and costs of feed of $1.91 per cwt. (each 75 percent that assumed in the United States), the increase in daily costs directly associated with adoption of rbST costs are $0.44. Thus, the use of rbST implies a loss of about $0.13 per day, or $12.74 over 98 day lactation. Although rbST achieves a very high percentage increase in milk yields, the absolute increase remains low because the initial base is so low. The cost of rbST is too high relative to the benefits achieved.[23]

The results reported in the Zimbabwe study were achieved on an experiment station where the quality of feed and management were high. The production increase achieved is substantially higher than the increment that would be achieved in most African dairies. As previously noted, use

TABLE 2.7 Incremental Revenues, Costs, and Profit per Animal from Use of rbST with Mashona and Nkone *Bos Indicus* Cattle in Zimbabwe

Incremental Revenue Assumptions:	
Milk price in $ per cwt.	$11.00
Milk price in $ per lb	$0.11
Average daily production per cow without rbST	1.00 lb
Average daily production per cow with rbST	3.78 lb
Percent daily response to rbST	278
Increase in daily production per cow with rbST	2.78 lb
Incremental revenue per day of rbST use:	
$0.11 per lb per day * 2.78 lbs =	$0.31
Incremental Cost Assumptions:	
rbST cost per cow per day	$0.36
Returns to mgmt.: included in net profit	$0.00
Milk Hauling, other non-feed variable costs,	
and increased feed costs: $2.82 per cwt.	
or $0.028 per lb: $0.028 per lb * 2.78 lbs =	$0.08
Total Incremental Costs per cow per day:	$0.44
Net Incremental Loss Per Cow Per Day	($0.13)
Net Incremental Loss Per Cow Per 98 day lactation	($12.74)

Source: Author's calculations, adapted from Marion and Wills, 1990, and Perrin, 1991.

of rbST in tropical conditions will require special management practices, including raising the nutritional quality of daily rations, feeding often to encourage intake, and providing additional shade. Use of rbST will challenge management in the developed countries, and the challenge will be much greater in developing countries, where the types of management skills required are traditionally much weaker. Providing adequate nutrition will also be difficult because the digestibility of tropical forage is low. The expected loss from applying rbST to *Bos indicus* cattle of this sort in Africa should be even greater than calculated above, because commercial farmers in Africa and other developing countries will probably achieve smaller responses to rbST than those assumed.[24]

Nonetheless, *Bos indicus* cows in many countries have milk yields that exceed the 150 kg for cows in the Zimbabwe study. Worldwide, milk yields from *Bos indicus* cattle range up to about 2,000 kg. Yields of about 1,000 are common, though this would be higher than the average (Cunningham and Syrstad, 1987). These higher base yields suggest some potential for using rbST on *Bos indicus* animals.

Some insight into this potential may be obtained from examining the effect of rbST on the *Bos indicus-Bos taurus* crossbred dairy cows which were also involved in rbST trials in Zimbabwe (Phipps et al., 1991). These crossbred cows have much higher milk yields than the *Bos indicus* cows. Moreover, crossbred animals make up an increasingly large fraction of dairy herds and account for a growing fraction of the milk produced in developing countries. Crossbreeds are the cattle whose response will likely determine whether rbST is applied in most developing countries.

The crossbred cattle utilized in Zimbabwe were administered rbST during a 16-week trial. These cows had relatively high milk yields for crossbred dairy cows in developing countries (Cunningham and Syrstad, 1987), having an average pre-treatment output per lactation of about 2,400 kg (5,280 lb). For these cows, application of rbST increased milk yields from 8.6 kg/day (18.9 lb) for the controls to 11 kg/day (24.2 lb) for treated cows, an increase of 2.4 kg/day (5.3 lb) or 28 percent. Although this percentage increase is somewhat higher than that which has been achieved on *Bos taurus* herds in developed countries, it is a smaller percentage increment than was achieved on the *Bos indicus* cattle in Zimbabwe. The decline in the percentage increment suggests that, at least for relatively low-yielding cattle breeds (or crossbreeds), the percentage increase in milk yields from use of rbST is likely to be inversely related to the initial yield.[25]

Nonetheless, for these cross-bred cattle, the high percentage gain on a reasonably high initial milk yield makes the use of rbST appear marginally profitable under the same assumptions regarding input and output prices used before.[26] See Table 2.8. Daily per cow profit is increased by roughly $0.07, and total profit per lactation is increased by $7.84 through use of rbST. The unit cost savings, using Perrin's approach, is calculated to be 2.7 percent.

This result demonstrates that administration of rbST to crossbred and perhaps even high yielding *Bos indicus* cows could be profitable under exceptional circumstances.[27] The crossbred animals in Zimbabwe, however, had unusually high pre-treatment milk yields and received high-quality experiment station management and nutrition throughout the trials. In addition to nitrogen-fertilized Bermuda grass pasture, each cow received a 3 kg fresh weight/day of concentrate supplement containing 200 g crude protein per kg of DM. Eight weeks after onset of the dry season, cows received a forage supplement of 3 kg DM/day of maize silage. Few other crossbred cows in developing countries will find such favorable circumstances.

Environment, nutrition and management in the Zimbabwe trials was extremely favorable relative to that in most developing country contexts. However, some developing country producers face production

TABLE 2.8   Incremental Revenues, Costs, and Profit Per Cow for Use of
rbST with Mashona/Nkone x Friesian Cross-bred Cattle in Zimbabwe

Incremental Revenue Assumptions:	
Milk price in $ per cwt.	$11.00
Annual production per cow without rbST	52 cwt.
Percent annual response to rbST	11
Calculations:	
Incremental milk production:	
52 cwt. * .11 =	5.7 cwt.
Incremental revenue:	
5.7 cwt. * $11.00 =	$62.70
Incremental Cost Assumptions:	
rbST = $.36 per cow per day	
Calculations:	
rbST price: $0.36 * 140 days use per cow	$50.40
Returns to mgmt.: included in net profit	0.00
Milk hauling: and other non-feed variable costs: $.65 * 5.7cwt.	3.71
Increased feed costs: $1.91 * 5.7 cwt.	10.89
Total incremental costs per cow per year	$65.00
Net incremental loss per cow per year from use of rbST	($2.30)

Source: Author's calculations, adapted from Marion and Wills, 1990, and Perrin,
1991.

environments which are more moderate and some of these producers may
find the higher potential profitability provided by an rbST price decrease
sufficient inducement to improve management and nutrition in ways that
will permit profitable adoption. The incentive to alter on-farm management
and nutrition in the required manner will increase if and when the price of
rbST declines. For example, if rbST were priced at $0.20 instead of $0.36
per daily cow dose, per cow profit would be increased by $17.90 per 112
day lactation. However, each of the examples presented assumes
parameters regarding environment, nutrition and herd management that
could not be met by most developing country producers.

It is worth remembering that even in California, where the profitability
of rbST calculated following the same approach appeared extremely high
for most farms, at the end of the first year of commercial availability, rbST
had been adopted only by about 20 percent of producers and was being
used on only about 8.5 percent of cows. In the foreseeable future, rbST use
is not likely to be used on more than a small proportion of developing

country dairy cows, perhaps 1-5 percent in regions such as Africa, and not more than 5-10 percent in the more favored milk producing regions such as Latin America or the Indian subcontinent. Moreover, in these regions the percentage saving in milk costs will be smaller than those in the United States. Applying plausible percentage milk increments to these figures, it is estimated that the use of rbST will achieve an increase in developing country milk production of only 1 percent to 2 percent over the next decade or so. However, adoption of rbST should increase over time as management, nutrition, infrastructure, marketing, and herd genetics continue to improve in these countries. Similarly, when the patents restricting manufacture of rbST end, the price of rbST should decline dramatically. The use of rbST could eventually become widespread in developing countries.

Notes

1. F1 cows demonstrate higher efficiency than second-generation (F2) and successive generations of crossbred animals.

2. In late 1994, the Economic Community extended the initial moratoria through the end of 1999. It now seems that the adoption of rbST in Western Europe will be delayed longer than anticipated when this manuscript was drafted.

3. Other management practices that are unassociated with rbST may also lead to higher incidence of udder infections.

4. Milk consumption in the United States rose slightly during the first year following introduction of rbST.

5. In Marion and Wills' farm decision model, adoption occurs only if rbST provides the owner of a cow herd with a minimum daily incremental return per cow, initially set at $0.40 per cow/day. Using a somewhat unusual approach, this minimum daily incremental return is treated as a cost. Their resulting comparison of the incremental revenues and costs associated with use of rbST shows a "profit." However, this "profit" is in addition to the return already incorporated in the cost structure. Thus, rbST is generally more profitable than it first appears.

6. Monsanto's optimal price may not yield a net profit. Monsanto experienced substantial delays in receiving permission to market rbST, its marketing expenses have been high, and several potential markets like Canada and Europe have declared moratoria on the use of rbST that will postpone sales well into the future. Consumers will benefit, however, when Monsanto's patent expires, the price of rbST declines, and the price of milk also falls.

7. Monsanto presumably was especially concerned with avoiding problems while rbST was being introduced that would have led to negative publicity .

8. Many producers have expressed considerable *ex-ante* reluctance to consider use of rbST, e.g., Saha, et al., 1994 and Klotz, et al., 1995. While reluctance to use rbST is partly based on fear of human and animal health hazards, it is also related to the economic parameters that affect the expected profitability of adoption. Farmers whose personal and farm characteristics seem likely to permit profitable adoption are more likely to consider adopting rbST. Wisconsin farmers have been identified in the press and by Monsanto as having resisted the introduction of rbST more than farmers in other states. The data on Wisconsin farm characteristics suggest that the use of rbST will be less profitable in that state than in California. Further, if adoption elsewhere does lead to a generalized decline in milk price, farmers with lower yields and smaller herds seem likely to lose.

9. Early information regarding rbST adoption indicates that many of the farmers who are adopting are using rbST on only that portion of their herds that are able to respond profitably. This practice reflects the heterogeneity of cow performance in many dairy herds. The practice will allow a larger number of dairy farmers to begin using rbST in the short run and should encourage farmers to improve herd genetics and management practices in order to gain the benefits of rbST on the rest of their herds.

10. Profitable use of rbST requires an efficient record keeping system to identify cow response to administration of rbST and to ensure that each cow's diet is appropriate to its response and condition.

11. Responses to rbST have been on the order of 18-19% in numerous trials, e.g., Burton, et al. 1990, McBride et al. 1990, Jordan et al. 1991.

12. Reed found that herds of a thousand cows are common in some areas of the state. Larger dairies provide workers with somewhat more training, and better worker training is associated with somewhat higher herd milk yields. Dairies with larger herds pay higher wages, provide workers with better benefits than do dairies with smaller herds, and thus probably have a higher level of management and labor capabilities, and greater chance of making profitable use of rbST.

13. Refer again to Table 2.1 for Marion and Wills' cost and revenue estimates. Perrin determines the unit cost of milk when rbST is used by dividing the total cost of cow maintenance, including the incremental costs associated with rbST use, by the total amount of milk produced when rbST is used. This unit cost is compared with the average cost of milk when rbSt is not used. While the estimates here suggest only a small reduction in unit costs, plausible changes in the assumptions result in larger reductions in unit costs, as noted in the previous paragraph and documented below. See also Table 2.2.

14. When the price of milk is $11.50 per cwt. and the price of rbST is $0.36 per daily cow dose, use of rbST allows a 5.9 percent reduction in the unit cost of milk.

15. Breeders have been able to select for response to many factors, including other drugs such as prostaglandins, and there is every reason to expect that breeders will be able to select for response to rbST as well.

16. Among those who tried and quit rbST, the high cost of rbST and problems with cow reproduction and with cow "burn out" were the main reasons for quitting. The same concerns were voiced by users who indicate that they plan to continue use. Users appear unconcerned with public opinion and seem to appreciate that their own use will have little effect on the price of milk (Butler, 1995).

17. Some Asian regions such as India have developed an intensive dairy sector based on the use of concentrates, but dairy sectors in Africa and Latin America are based primarily on pasture. Latin America's tropical region is responsible for about 30 percent of all milk produced in the developing countries, and most of this milk is produced from pastures.

18. To make the use of these animals profitable, the farmer needs to obtain a number of different products, including traction. However, animals that are adapted biologically to relatively harsh conditions and to a low level of management inputs, and that are genetically suitable for producing several different outputs, are usually incapable of providing a high level of output of only one product, such as milk, even if such specialization is desired (Vercoe and Frisch, 1980; Preston, 1976 and 1977; Jarvis, 1982).

19. Trypanosomiasis, also known as sleeping sickness, is a chronic, sometimes epidemic disease affecting cattle, humans, and other animals. It is caused by a protozoan blood parasite, genus *trypanosoma*.

20. The 500 mg dose rate used was the rate recommended for *Bos taurus* cattle; further trials with alternative doses might achieve a higher response from *Bos indicus* cattle.

21. No figures are reported in the study for total milk yield from the control cows, so the pre-treatment yields are only approximate.

22. Phipps et al. (1991) found that rbST had little impact on milk yields when it was applied late in the lactation and noted the need for research to determine the optimum timing for treatment initiation.

23. Using this analytical framework, rbST appears marginally profitable if the price of rbST is reduced to $0.20 per daily cow dose, assuming the other assumptions held. However, such a result should not suggest that the use of rbST on such cattle is imminent, or ever likely.

24. In the Zimbabwe Holstein trials, cows were divided into three groups based on state of lactation: 90-120, 121-150, and 151-180 days. Milk yields increased over the controls during the period of treatment by 13, 13, and 0 percent, respectively, again indicating that application of rbST during the latter stages of lactation has little effect. The trials for Friesians and Jerseys showed analogous increases of 19 and 15%, respectively. Each of these production increments is similar in magnitude to those that have been achieved in developed countries when rbST has been applied to similar cattle. That similarity is another indication that the production environment within which the trials took place was very favorable, and thus marketedly more favorable than the environment faced by most developing country dairy producers.

25. Moreover, administration of rbST to the Zimbabwe crossbred cows achieved an increase of only about 15% for the whole lactation. The lactation for these cattle was relatively short and relatively high yields were achieved early in the lactation before rbST was administered. Thus, the percentage increase in total milk production is less than half that achieved during the period of administration.

26. Cows with higher base yields are likely to achieve higher absolute increments from treatment with rbST and it is the absolute increment, not the percentage increment that is economically important.

27. The Zimbabwe data indicate that the use of rbST can be profitable for cattle breeds whose top level of production falls far short of that achieved by the specialized dairy breeds used in developed countries, provided the absolute production increment is high and the input-output price relationships are favorable. This will not happen frequently.

3

Multiple Ovulation and Embryo Transfer (MOET)

The second biotechnology considered in this study is multiple ovulation and embryo transfer (MOET). MOET involves the use of drugs to stimulate multiple ovulation, followed by the retrieval and transfer of the resulting embryos from a donor cow to as many recipient cows as are necessary for completion of the pregnancies (Seidel and Seidel, 1981). MOET has become an important commercial technology in the dairy industry during the last two decades (Greve, 1986; Ruane, 1988). It is carried out predominantly in the developed countries, although a large number of developing countries are now undertaking MOET at an experimental level.

Every cow has perhaps 75,000 potential eggs in her ovaries, but most give birth to only about 4 calves over a lifetime. During the past 25 years, techniques have been developed to superovulate cows and recover (non-surgically) a large number of embryos at a time. The use of MOET allows a donor cow to produce about 25 calves over her life instead of about 4, and this number may rise to over 100 (Cunningham, 1988).

The main purpose of MOET is to allow for an increase in the intensity of genetic selection. Genetic selection is the most important mechanism by which animal performance has been improved over time. The principle of genetic selection is simple. Animals demonstrate a natural variation with respect to numerous characteristics such as the ability to achieve higher milk yields or higher calving rates under specified conditions. Such characteristics are significantly hereditary. The genetic potential of the herd to express the desired characteristics can be increased by culling those animals whose performance is particularly poor, and by giving preference in breeding to those whose performance is particularly good.

Artificial insemination has been a powerful tool to increase the intensity of genetic selection in breeding schemes in the developed countries when combined with the analysis of the milk production records of the bulls' daughters. Given that one bull can sire many animals, artificial insemination has allowed breeding to be restricted to the top 1 percent of bulls whose offspring have demonstrated the highest milk productivity, leading to a higher rate of herd genetic improvement. The performance characteristics of different bulls can be evaluated through the testing of their progeny, thus providing greater precision regarding the heritability of traits under actual field conditions. Semen from those bulls whose progeny are highly productive is then selected for widespread use in artificial insemination.

Cow selection is traditionally much less intense than bull selection. Cows that have low milk yields or reproductive problems are culled from the herd. However, the proportion of cows that can be culled each year is usually limited, depending on the number of heifers available for herd replacement. Because cows produce fewer than one calf a year, on average, half of which are male, relatively little selection is possible in conventional breeding systems. Seventy to 90 percent of heifer calves must be raised as replacements for those animals that are culled for reasons other than low production. With so few offspring from each cow, there is less opportunity to evaluate the performance of the progeny coming from cows than from bulls. It is thus more difficult to determine the heritability of desirable traits. With MOET, however, the number of progeny rises significantly.

MOET can be utilized as a stand-alone breeding technology or, preferably, in combination with artificial insemination. A higher intensity of selection can be achieved among males using artificial insemination than can be achieved among females using MOET. MOET may have to be combined with other technologies in developing countries if it is to be economical. This issue is discussed below.

MOET may also be used to achieve other goals in developing countries. For example, MOET may be used in crossbreeding dairy cattle to produce a continuous supply of first generation (F1) crossbred cows. This approach has potential to achieve higher milk yields as a result of the higher heterosis in first—as opposed to successive—generation cows having the same genetic mix.[1] MOET could also permit rapid expansion of the rather small herds of special animal breeds (such as the relatively trypanosomiasis-tolerant N'dama), if such breeds are found to be particularly valuable for specific uses in developing countries. Finally, MOET might be used, through the storage of frozen embryos, to preserve the germplasm available in local cattle breeds before these genes are irretrievably lost through cross-breeding with exotic livestock.

Scientific advances are making the MOET technique increasingly economical. The freezing and thawing of embryos are now routine, with little effect on embryo viability. Embryos can also be transferred to recipients non-surgically, with transfers carried out by a team on the farm. The development of "in vitro" fertilization has reduced the cost of embryos and the pre-sexing of embryos will allow choice of the desired sex of offspring. The latter is of considerable value in cases where there is a significant difference in the value of male and female calves, as is usually the case with dairy animals.

Nonetheless, MOET is a less powerful genetic tool than artificial insemination. Genetically-superior donor cows cannot be identified with nearly the accuracy of genetically-superior bulls. Furthermore, once a genetically-superior female has been identified, the total number of offspring that can be obtained from that cow remains much smaller than the number that can be obtained from a bull. The number of offspring that can be produced via MOET alone are so small as to have barely measurable influence on the general population (Wideman, 1982).[2] However, faster genetic progress can be made by combining MOET and artificial insemination than by using artificial insemination alone. Table 3.1 shows the effect of increasing the number of progeny per cow on herd selection intensity and on progeny genetic superiority.

TABLE 3.1   Effect of Selection on Increasing the Number of Progeny Per Cow

No. of progeny per cow per lifetime	Percent of cows selected as dams of replacements	Genetic superiority of progeny (% of average)
4	59.0	2.15
8	29.5	3.80
20	14.8	5.07
40	7.4	6.14
80	3.7	7.15

Source: Wideman, 1982.

The Expected Impact of MOET on Milk Production in Developed Countries

MOET is expected to be of value mainly when it is utilized as an integral component of a broader, well-planned and implemented breeding program (e.g. Stranden, et al., 1991). Highly organized artificial insemination-based selection programs in developed countries are achieving increases in milk yields equal to about 1 percent per year (Norman and Powell, 1986; Dentine and McDaniel, 1987; Pearson and Freeman, 1973; P. Cunningham, personal correspondence). Most analyses suggest that using MOET with artificial insemination could increase the rate of genetic improvement by about .3 percent above the levels achieved with artificial insemination alone. Thus, in the breeding systems currently practiced, the use of MOET might increase the rate of gain to 1.3 percent per year, providing a net gain over a decade in herd genetic quality of slightly more than 3 percent. As the use of MOET is still not fully refined even in developed countries, it is assumed in this study that, where the use of MOET becomes economically profitable, MOET will increase the rate of improvement in herd genetic potential by about two percent per decade. See Chapter 4.

Although the expected gain from use of MOET is significant, under many circumstances the relatively high costs of MOET will not justify the extra genetic gain (Seidel, 1986). Nonetheless, the potential gains from MOET are higher than those currently being achieved. For example, the theoretical rate of gain from artificial insemination in an efficient system is about 2 percent per year without using MOET, and perhaps 2.3 percent with MOET. The theoretical rates achievable if MOET is integrated into an Open Nucleus Breeding System (ONBS), as discussed below, are as much as 4.8 percent annually (Smith 1988a). The ONBS system is just now being implemented at an experimental level in developed countries. However, an ONBS has the potential to interact positively with MOET, particularly as molecular genetics advances, increasing significantly the rate of herd genetic progress. This effect is probably at least a decade away.[3]

The Expected Impact of MOET on Milk Production in Developing Countries

MOET has been suggested for several uses in developing countries. Three possible uses are examined here. The first two involve efforts to increase the rate of herd genetic progress through, one, combining MOET with artificial insemination and field testing in well-planned dairy breeding programs, and, two, using MOET as a component of an ONBS in order to avoid the need for field testing. The third approach uses MOET to produce a continuous supply of first generation (F1) crossbred cows as a means of sustaining higher milk yields.

MOET and Artificial Insemination for Progeny Testing

The use of MOET will face serious challenge in developing countries. Although non-surgical embryo transfers are more simple, cheaper, and involve less risk than surgical transfers, even the use of non-surgical embryo transfers in developing countries faces many obstacles. The disease, parasite and climate challenge, the low level of animal nutrition, the low quality of on-farm management and labor skills, and inferior communication and transportation infrastructure will increase costs and reduce the proportion of successful pregnancies achieved by MOET relative to those achieved in developed countries. This will increase the cost per successful embryo transfer. These factors will also lower the value of the progeny achieved, both by lowering their physical productivity and by reducing the value of their output, e.g., because of less favorable input-output price relationships.

Logistics and management are the most serious constraints, if the history of artificial insemination is a guide. The benefits of artificial insemination depend crucially on identifying those bulls whose offspring produce significantly higher milk, and then successfully inseminating dairy cows with semen from the preferred bulls. It is widely agreed that artificial insemination has functioned poorly in most developing countries. It has been impossible in most environments to maintain an adequate field recording system—without which the process of genetic selection is much less accurate (Cunningham, 1989a; Vaccaro, 1994). Important genotype-environment interactions can also lead to significant changes in the ranking of sires across different environments (Buvendran and Petersen, 1980; Blake et al., 1988; Blake et al., 1989; and Holmann et al., 1990b). The infrastructure available has been insufficient for artificial insemination to function—resulting in low calving rates. The infrastructure and management problems that the use of MOET will confront in developing countries will be even greater than those encountered by artificial insemination, and the field recording requirements will be similar. Although MOET may be profitable in special circumstances in a few developing countries, it will not soon become a generalized practice and therefore will not significantly affect the genetic potential of developing country dairy herds unless combined with additional technologies.

MOET and the Open Nucleus Breeding System (ONBS)

A second approach utilizing MOET to achieve more rapid genetic selection in developing countries involves a small open nucleus breeding herd. The Open Nucleus Breeding System (Nicholas and Smith, 1983; Smith, 1988a) concept is relatively simple. A genetically-superior nucleus herd of perhaps 250 cows is established under controlled conditions to permit testing and genetic selection. Development of an ONBS begins when outstanding females are selected for inclusion in the ONBS herd by screening the base population (e.g., herds in the surrounding region). Genetic selection via natural service or artificial insemination is used to achieve progeny with higher milk yields, and the superior progeny created in the nucleus herd are then released for use in the national herd, having a direct effect on both milk output and on genetic improvement. Use of an

ONBS avoids the need for field testing, and also alleviates the problem associated with genotype-environment interactions and infrastructure. By developing improved animals whose performance within the local environment can be better predicted, the use of MOET can reduce production risks.[4]

Although implementation of an ONBS does not require MOET, use of MOET can theoretically increase the rate of genetic improvement achieved from an ONBS. MOET, along with superior sires, can be used to obtain embryos from the elite females within the ONBS herd. These embryos can be transferred to and carried by recipient cows from the base population. The males in the resulting offspring can then be evaluated using the performance of their siblings and half-siblings, with the best males used in artificial insemination or natural service in the base population. In theory, the ONBS can be used to improve purebred and stabilized crossbred genotypes. Use of an ONBS can also limit MOET activities to a central unit, with better control of animal husbandry in the ONBS itself. Completion of all MOET activities within the ONBS limits should facilitate development of a well-trained staff, and simplify logistics so that the breeding system can be more easily and more completely controlled (Smith, 1988b).[5]

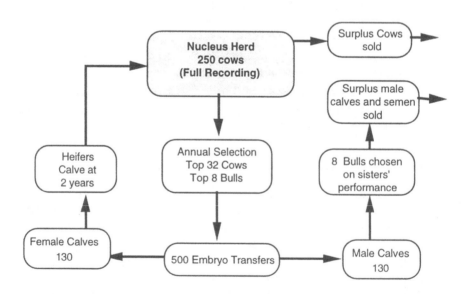

Figure 3.1 Example of Open Nucleus Breeding System for Dairy Cattle

Source: Hodges, 1988.

Success of the ONBS is closely related to the degree of genetic selection possible, as shown in Figure 3.1. In this example, the 32 top cows from the herd of 250 cows are selected for MOET with 8 sires. Assuming recovery of 16 embryos per cow, 512 in total, and a 50 percent pregnancy rate, 260 calves are expected. Of these, approximately 8 will result from each cow, 4 of each sex. Each of the 130 male calves are then evaluated as bulls through the performance of their 4 full sisters and 12 half sisters. The best 120 of the older cows are retained each year, with the other 130 distributed to the base population when replaced by the 130 heifers who enter for their first lactation (Hodges, 1988).

Nonetheless, even in the ONBS framework, the number of embryos recovered and the number of successful pregnancies achieved through MOET is likely to be substantially lower in a developing country environment than in a developed country environment. Cows in lowland tropical areas have lower reproductive efficiency even when managed well (Vaccaro, 1991). As a result, the average number of embryos recovered may be much lower, perhaps 4 instead of 16, and the average number of successful pregnancies may be fewer, perhaps 25 percent instead of 50 percent. These reductions, if experienced, would dramatically affect the efficiency of the ONBS. Following the same example, to achieve the same number of embryos under these conditions, the number of cows used for MOET would rise to 128, approximately half the herd, allowing for much less selection. Alternatively, the size of the herd could be increased, but at substantial cost, or the number of calves reduced—with fewer bulls distributed to the base population and fewer siblings available for evaluation. The problem is exacerbated if the percentage of successful pregnancies also declines.

The ONBS also will function only if farmers in the local population are willing to sell their very best animals to the ONBS, and to utilize the resulting offspring in return.[6] If the best animals are not obtained, the herd would begin with a lower genetic potential. Even if the use of the ONBS were to achieve rapid genetic improvement, if the ONBS herd starts from a lower level, the effort would require a long time to have a positive impact on the genetic potential of the national herd. Obtaining truly superior cows from local farmers might be manageable in areas such as those in India where many small-scale farmers already purchase most of their heifers from a central, often public breeding institution rather than breed and raise their own replacements. The problems could be less easily surmountable in Latin America and Africa where producers are accustomed to breeding their own animals and could be suspicious of any centralized institution.

Milking practices common to most developing countries contribute to a high variability in cow milk yields.[7] Observed cow milk yields vary significantly from day to day, and even over longer periods (Vaccaro, 1988).

This variation makes the recording of sibling and half-sibling milk output less precise and can reduce the efficiency of selection. If animals are to be distributed to the base population for use by local farmers, they must be evaluated within a similar context. At the nucleus level, reasonable standardization can be achieved to facilitate between cow comparison, and appropriate analytical techniques can be used to process the data.

Finally, to have significant effect on a national dairy herd, the ONBS must be very large or many ONBSs must be established. Thus, this approach requires substantial capital, a large trained staff, and the creation of significant logistical infrastructure. For example, assuming that an ONBS could provide 130 bull calves each year, and that each bull calf remained in service an average of 5 years and was capable in natural service of mating with 25 cows annually, at full development the ONBS would affect the output of 16,250 cows. Even if one focused the output of an ONBS only on those farms having the 20 percent top yielding cows in an "average" Latin American country, for example, an ONBS would not be capable of affecting the output of more than about 10 percent of such cows. Moreover, the number of cows affected would be significantly smaller than as calculated above if the reproductive efficiency of the ONBS center is lower than assumed.

These problems are well understood by those who have proposed ONBS for consideration in developing countries (Smith, 1988b). There is as yet no data to indicate how the ONBS will operate in practice in developing countries, nor under what conditions combining it with MOET will be economical. The use of MOET within an ONBS offers significant promise for the longer run (Kasonta and Nitter, 1990), but it will probably have little perceptible impact on milk production in the developing countries during the next decade.

MOET and the Continuous Production of
First Generation (F1) Crossbred Calves

MOET might also be utilized along with simple crossbreeding to increase the genetic potential of dairy cows. Preliminary analysis indicated that this approach was the most likely to allow profitable use of MOET in developing countries in the intermediate future. As a result, substantial detail is provided.

Before discussing this use of MOET, we first discuss the possible benefits of crossbreeding, which are an integral part of this approach. The low milk yields observed in most developing countries are partly due to a historical lack of selection for milk production. Genetic selection among dairy cattle in the existing herd can and will raise milk productivity, but such a process allows only for a rather slow increase and there are many obstacles toward conventional selection approaches, as have been discussed. A more rapid approach toward genetic improvement involves the introduction of genes from animals demonstrating higher milk productivity, e.g., crossbreeding *Bos taurus* breeds with the local *Bos indicus* breeds.

Substitution of *Bos taurus* for *Bos indicus* breeds is also an alternative. However, the importation of *Bos taurus* milk breeds from temperate regions has been widely unsuccessful (Brumby, 1979). Such animals require

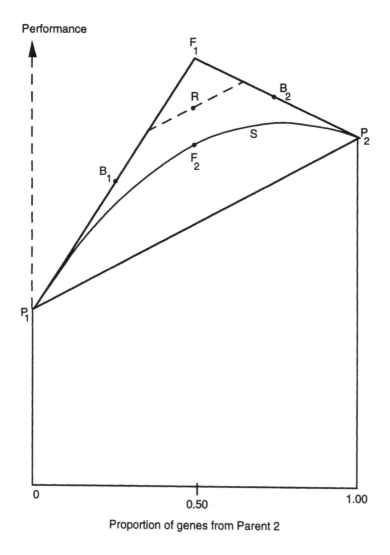

Figure 3.2 The Greek Temple Model: Expected performance in Rotational (R) and Synthetic (S) Breeding Systems

Source: Cunningham and Syrstad, 1987.

conditions that are not present in most tropical regions. They are unable to function in hot, humid regions, where disease and parasite threat is great and where the nutritional regime is low. The performance of such animals is poor, usually well below that displayed by the local animals that they are intended to replace (Cunningham and Syrstad, 1987; Vaccaro, 1994; Turton, 1981; Madalena, 1981).

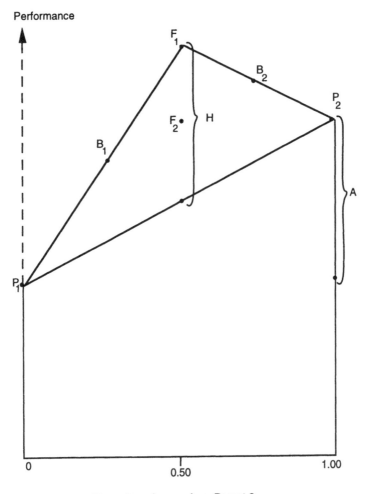

Figure 3.3 The Greek Temple Model: Expected Performance of Parental (P_1 and P_2), F_1, F_2 and Backcross (B_1 and B_2) Groups as Functions of Additive (A) and Heterosis (H) Effects

Source: Cunningham and Syrstad, 1987.

In contrast, crossbreeding *Bos indicus* and *Bos taurus* types has been shown to result in animals that yield significantly more milk than either purebred in tropical conditions. Crosses combine some of the milk producing ability of the temperate *Bos taurus* breeds with adequate climatic adaptability from the tropical breeds. Typically, such crossbreeds produce more than double the amount of milk per lactation of the local breeds (Cunningham and Syrstad, 1987; Turton, 1981 and 1985; Vaccaro, 1994). This increase is partly due to the combination of the attributes of both parents, and partly to hybrid vigor which is specific to the first (or F1) cross. When these first crosses are interbred or backcrossed to either parent breed, part of the merit of the first cross is usually lost.

The primary theoretical effects of crossbreeding are shown in the Greek Temple Model developed by Cunningham (1987). The performance (e.g., milk yield) of different parent breeds, P_1 and P_2, relative to those of the first generation cross, F1, the second generation cross, F2 (the cross of two F1 animals), and the backcrosses (crossing an F1 animal with a purebred) is shown in Figure 3.2. As noted, the difference in the mean yields of the parent stock is the additive effect, while the increase in performance (e.g., average milk yield) manifested by the crossbreed relative to the average output of the parent breeds is known as heterosis.

One of the striking aspects of crossbreeding is the potential for the crossbreed to perform better than either of the parent breeds in a given environment, given the potentially positive effect of heterosis. As a result, it is often possible in tropical areas to substantially increase herd performance through crossbreeding much more rapidly than could be achieved by selection only.[8] Indeed, there is substantial evidence that crossbred animals regularly produce yields that are significantly higher than those of the local cattle population (Cunningham and Syrstad, 1987; Vaccaro, 1979 and 1994; Turton, 1981 and 1985; Madalena, 1981). Thus, crossbreeding, whether through natural or artificial insemination, is often a means of achieving rapid increases in milk yield.

Nonetheless, heterosis is highest for the first generation. The subsequent generation (F2) retains only about 50 percent of this effect and F2 cow milk yields are usually significantly lower than those of the F1 cow.[9] The decline in performance can be partially avoided by the use of rotational breeding schemes that involve continual crossing of purebreds with alternate generations. Successive generations then contain a varying percentage of local and exotic genes. The performance of the F1 generation relative to the F2 and other generations (obtained through a rotational breeding strategy) is shown in Figure 3.3. Also shown is the expected performance of a synthetic breed (obtained from continued *inter se* matings) with varying percentages of local and exotic genes. As noted, the F1 generation manifests a significant advantage over others.

Data on the actual advantage is available from Cunningham and Syrstad (1987), who assembled information on the relative productivities of F1 and F2 generation crossbred dairy cattle in five countries: India, Sri Lanka, Brazil, the United States, and Australia. Although these data are generated from experiments that are not always directly comparable, F1 generation animals are systematically more productive in each country except Australia. Moreover, the F1 animals appear more productive in terms of each of the major characteristics considered: age at first calving, milk yield (over 300 days and/or over the whole lactation), and calving interval. Most of the benefits of the F1 generation appear in milk yields.[10] This difference seems to be on the order of 500 kg to 600 kg in the Indian subcontinent and somewhat greater in the U.S., although the differential is negative in Australia. See Table 3.2. The averages shown are simple averages, with no attention to the number of cows involved in the study, nor to the region considered.[11] Based on more limited information, the age at first calving was about 10 percent younger and the calving interval was about 7 percent shorter for F1 generation cows.

Since F1 generation animals demonstrate significantly higher productivity than F2 generation animals within the same developing country production environment, continuous production of F1 generation animals is potentially valuable. Thus, contrary to the case of exotic animals, which require significant environment-improving investments lest they perform absolutely poorer than the animals they are intended to replace, the F1 systematically performs better than the F2 with little or no change in the

TABLE 3.2   Yield Difference Between F1 and F2 Generations

Country	Milk Yield (avg. F2 lactation (kg)	F1-F2 Yield Differential (kg)	% Increase F2 to F1
India	1,833	576	26
Sri Lanka	981	567	47
Brazil*	2,482	395	13
United States	1,712	648	38
Australia	674	−155	−27
Simple Average	1,536	406	19
Simple Average of LDCs	1,765	513	29
Simple Average of Indian Subcontinent	1,407	572	37

* (Pigangueras breed; see text)

Source: Author's calculations from data in Cunningham and Syrstad, 1987.

environment. It is impossible, of course, to produce a continuous flow of F1 animals with natural service or artificial insemination, but MOET could be used to achieve a continuous flow of F1 animals. The question is whether the benefits are worth the costs.

A sense of the magnitude of the benefits from producing a continuous flow of F1 generation animals can be obtained from data in Cunningham and Syrstad (1987). They calculated the heterosis observed in F1 animals in the reported trials, finding substantial heterosis for all traits except lactation length. For example, heterosis was 28 percent for milk yields, and 37 percent when milk yield was expressed per day of calving yield. As they note, assuming that only additive and dominance effects prevail, the milk yield displayed by F2 animals, relative to the mid-parent mean, should be half that of the F1 animals, e.g., 14 percent above the mid-parent mean for milk yield.

Observations on the performance displayed by F2 animals were generally consistent with this prediction except for milk yield, where performance was significantly lower. Many F2 animals achieved milk yields that were about equal to the mid-parent mean. Several explanations have been offered for the failure of F2 generation cows to achieve the expected milk yields (Cunningham and Syrstad, 1987), but none has proved definitive. Regardless, the differential productivity between F1 and F2 generations is larger than expected.

Somewhat surprisingly, given the expectation that higher heterosis should cause a higher absolute increase when the base yield is higher, the absolute increase in the F1-F2 milk yield differential shown in the experimental data is almost constant, irrespective of the base yield of the F2 generation. For example, in India, although F2 generation animals yielded 1833 kg per lactation, the F1-F2 differential was almost the same as that manifested by animals in Sri Lanka whose F2 base milk yield was only 981 kg. It seems that this could not easily result from a systematic difference in the quality of management, since this would imply that where management was adequate to achieve only half the basic yield, the effect of heterosis on yields was unchanged.

Cunningham and Syrstad (1987) report no data on the F1-F2 differential for Latin America or Africa. They report data from experiments in several other countries that indicate that heterosis was present in the F1 generations, but no data were available regarding the decrease in heterosis in the F2 generation. However, if F2 animals manifest no heterosis in terms of milk yields—an assumption broadly consistent with the reported data, the F1-F2 differential may be approximated by simply comparing F1 generation yields with the mean yield of the two parent breeds. The average F1-F2 differential estimated by this indirect method is similar in absolute magnitude and in percentage terms to the average F1-F2 differential derived when information on F2 generation milk yields are reported in Cunningham and Syrstad (1987).[12] See Table 3.3. It appears that the F1-F2 differential is of broadly similar magnitude on all continents, though there may be economically important variations in the differential among breeds and local context.[13]

Having characterized the magnitude of the F1-F2 differential, consider next the operational problems associated with using MOET to produce a continuous supply of F1 female calves in developing countries. Two methods could be used. First, MOET could be used to superovulate a local cow (donor), collect the multiple eggs so obtained, fertilize them with semen imported from *Bos taurus* bulls, and transfer the F1 embryos to recipient local cows, which, irrespective of generation, will bear an F1 female calf.[14] This approach could be repeated each year, thus maintaining a dairy herd of F1 generation cows.

Second, ova might be collected from *Bos taurus* donor cows in developed countries, fertilized with semen from *Bos indicus* bulls, and the resulting embryos transported to the developing country for transfer into local

TABLE 3.3   Productivity Difference Between F1 and F2 Generations
(Additional Observations)

Country	Milk Yield (avg. lactation for F1: kg)	% Yield Increase Over Midparent mean	Absolute Increase (kg)
Egypt	2112	34	542
Kenya*	1523	3	44
Iraq	3070	21	526
Nigeria	1690	0	(5)
Colombia	2000	47	516
Costa Rica	1780*	21	359
Simple Average	2029	21	397

* A simple average of the results from two reported trials.

+ In the experiments compiled by Cunningham and Syrstad, 1987, the F2 generation cows have the same milk yields as the midparent mean, implying no heterosis for the F2 generation. Assuming that F1 heterosis equals the F1-F2 differential , the F2 generation manifests no heterosis. Yield data from F1 generation cows in several countries can be used to calculate the expected F1-F2 yield differential.

Source: Author's calculations from data in Cunningham and Syrstad, 1987.

recipient cows. This approach could be less costly, either because the superovulation and collection of ova could be carried out more efficiently in a developing country, or because "generic" embryos could be produced more cheaply from the ovaries collected at slaughterhouses. American Breeders Service has developed a method to recover eggs from ovaries collected at slaughterhouses. Although the pedigree of the slaughtered cows is indeterminate, the average milk yield of U.S. dairy cattle is very high. Theoretically, all such cows have sufficient genetic potential to allow for a significant increase in milk yield when crossed with semen from *Bos Indicus* bulls from developing countries, and the "generic" embryos so produced would be cheap.

The Expected Value of Using MOET to Produce an F1 Calf

The economic gain that can be achieved via the use of MOET to obtain an F1 instead of an F2 calf is modeled as follows. Assume that a dairy producer has an F1 cow that is to be bred and the producer must choose between using MOET and natural service. If MOET is used, the producer obtains an F1 calf with probability alpha, and otherwise nothing, while if natural service is used, an F2 calf is obtained with probability beta, and otherwise nothing. The main factors determining which choice is more profitable are:

1. the price of milk,
2. the price of feed,
3. the price of other inputs required to utilize MOET instead of natural service or artificial insemination,
4. the differential milk yield between F1 and F2 generations,
5. the differential fertility between F1 and F2 generations, as reflected in the age of first calving, calving interval, and productive life,
6. the differential value of calves from a F1 and F2 generation cow, if any, including the differential value of male and female calves and their respective probabilities if MOET allows for sexing of embryos,
7. the interest rate used to discount future benefits and costs,
8. the cost of natural service or artificial insemination, versus that of MOET.

These factors are incorporated into Equations 3.1-3.10 to determine the profitability of using MOET to produce an F1 generation calf out of an F1 cow, relative to that of using natural service or artificial insemination to produce an F2 generation calf out of an F1 cow.

Equations 3.1 and 3.2 calculate the expected profitability of using MOET and natural service or artificial insemination, respectively, to achieve a pregnancy from the same F1 cow.[15] The dairy producer must pay the cost of MOET or natural service, which produces a calf nine months later with a given probability that depends on the method used.

3.1. $\pi_{MOET} = \alpha \left(V^{F1} / (1+r') \right) - C_{MOET}$

3.2. $\pi_{NS} = \beta \left(V^{F2} / (1+r') \right) - C_{NS}$, where

π_{MOET} = the expected present value associated with using MOET one time,

π_{NS} = the expected present value associated with using natural service one time,

α = the expected probability of a successful pregnancy when MOET is used,

β = the expected probability of a successful pregnancy when natural service is used,

C_{MOET} = the cost of MOET,

C_{NS} = the cost of natural service,

r' = the nine month real interest rate.

The potential gain from a successful pregnancy is usually high relative to the cost of either MOET or natural service. Thus, there is a high expected gain from repeating MOET or natural service if the first try does not produce a pregnancy, as is likely since the expected probability of each try is rather low. Indeed, farmers will usually find it profitable to repeat natural service or MOET several times before abandoning the effort to obtain a calf.[16] The expected value of a plan to try a fixed number of times before stopping can be seen as the expected value of adopting a given breeding strategy. Use of such a strategy follows the rationale that if it is profitable to use one method instead of another one time, it remains profitable to repeat the use of the more profitable method.[17] For example, the sum of the terms in Equations 3.3 and 3.4 shows the respective expected present values of repeating MOET and natural service four times.

3.3. $\Pi_{MOET} = \pi_{MOET} + (1-\alpha) \pi_{MOET} + (1-\alpha)^2 \pi_{MOET} + (1-\alpha)^3 \pi_{MOET}$

3.4. $\Pi_{NS} = \pi_{NS} + (1-\beta) \pi_{ns} + (1-\beta)^2 \pi_{ns} + (1-\beta)^3 \pi_{ns}$, where

Π_{MOET} = the expected present value associated with using MOET up to four successive times,

Π_{NS} = the expected present value associated with using natural service up to four successive times.

The expected values of using MOET and natural service are related inversely to their respective one-time probabilities of success and to their costs, and directly to the value at birth of F1 and F2 calves. Paradoxically, the ranking of the MOET and natural service that is obtained from comparing the expected values of one try of each approach can differ from the ranking that is obtained from comparing the expected values of a set of tries. This is caused by the negative association between the probability of success on the first try and the probabilities that a future try is needed.[18]

The values at birth of the F1 and F2 calves, which are used in Equations 3.1-3.4, are derived in Equations 3.5-3.10.[19] In Equations 3.5 and 3.6, the values of F1 and F2 calves are derived as simple averages of the values of male and female calves, assuming equal probabilities of occurrence. Male calf values are assigned arbitrarily, as explained later. Female calf values are derived as the expected present value of the respective calves' future production, less costs, as shown in equations 3.7 and 3.8. This formulation allows for recursive effects so that changes in milk and feed prices affecting the value of the female calf both directly and also indirectly via the value of its own future calf production.

3.5. $V^{F1} = (V_f^{F1} + V_m^{F1})/2$

3.6. $V^{F2} = (V_f^{F2} + V_m^{F2})/2$

3.7. $V_f^{F1} = \theta \sum\limits_{i=1}^{8} \dfrac{P_i M_i^{F1} + V_i^{F1} - \tilde{C}_i^{F1}}{(1+r)^i} - \sum\limits_{i=1}^{8} \dfrac{C_i^{F1}}{(1+r)^i}$

3.8. $V_f^{F2} = \phi \sum\limits_{i=1}^{7} \dfrac{P_i M_i^{F2} + V_i^{F2} - \tilde{C}_i^{F2}}{(1+r)^i} - \sum\limits_{i=1}^{7} \dfrac{C_i^{F2}}{(1+r)^i}$, where

V^{F1} = the expected value at birth of a F1 generation calf,

V^{F2} = the expected value at birth of a F2 generation calf,

V_f^{F1} = the value at birth of a female F1 generation calf,

V_m^{F1} = the value at birth of a male F1 generation calf,

V_f^{F2} = the value at birth of a female F2 generation calf, and

V_m^{F2} = the value at birth of a male F2 generation calf.

P = the price of milk,

M_i^{F1} = F1 milk production during one lactation,

M_i^{F2} = F2 milk production during one lactation,

C^{F1} = fixed feed and maintenance costs for F1 generation cow,

C^{F2} = fixed feed and maintenance costs for F2 generation cow,

$\tilde{C}^{F1}$ = feed and maintenance costs that vary with milk production for F1 generation cow,

$\tilde{C}^{F2}$ = feed and maintenance costs that vary with milk production for F2 generation cow,

i = year subscript,

θ = probability of a successful pregnancy after four tries with embryo transfer, and

φ = probability of a successful pregnancy after four tries with natural service.

r = the annual real interest rate.

The present value of an F1 female calf depends primarily on the value of the milk and the calves that it will produce during the mature period of its life, less its cost of maintenance which is assumed to vary with the level of output.[20] A discount rate of 10 percent is assumed.

F1 cows have somewhat higher fertility than F2 cows, although the differential reflected in the data collected by Cunningham and Syrstad (1987) is not large. This fertility differential allows the F1 cow to calve several months earlier when natural service is used, providing an earlier first lactation as well. The F1 cow is also likely to have shorter calving intervals and, as a result, may bear one or more additional calves and achieve one or more associated additional lactations during its life. The same benefit might occur if F1 cows are used as recipients of embryos. To capture this effect, each F1 cow is assumed able to produce a calf in years 4-8, a total of 5 calves and lactations, while each F2 calf is expected to be able to produce a calf in years 4-7, a total of four calves.

Equations 3.9 and 3.10 could be altered to achieve a more precise fertility differential, but the formulation shown provides a reasonable approximation of the effects of expected fertility differences between F1 and F2 cows. Moreover, the purpose of this exercise cannot be to indicate definitively whether it is profitable to use MOET to produce F1 cows, but rather to determine the likelihood under some fairly general assumptions that it might be profitable and to indicate the factors which are most important in determining that outcome.

The probability that a cow has a calf during each year of its mature life is related to the breeding method used. Separate probabilities are calculated for MOET as shown in Equations 3.9 and 3.10, assuming four tries each for the use of MOET and natural service, respectively.

3.9. $\theta = \alpha + \alpha(1-\alpha) + \alpha(1-\alpha)^2 + \alpha(1-\alpha)^3$

and

3.10. $\phi = \beta + \beta(1-\beta) + \beta(1-\beta)^2 + \beta(1-\beta)^3$

In Equations 3.7 and 3.8, a change in the assumed probability of success affects the expected value of MOET both directly, through α in Equation 3.3 and also indirectly, through θ in Equation 3.7.

The F1 and F2 female calves that are created by MOET and and by natural service, respectively (See Equations 3.5 and 3.6), are assumed to have equal value in the calculations carried out subsequently. If MOET is used to breed the calves of calves, there should be no differential in their value. If natural service is used, F2 and F3 calves will result. F3 calves are somewhat more productive, though the difference between F2 and F2 calves is smaller than between F1 and F2 calves (Cunningham and Syrstad, 1987).[21]

If natural service is used, the resulting F2 calf is male with 50 percent probability. If MOET is used, the same is true unless the embryo is sexed, which is now possible at added cost. That male and female crossbred calves have different values, can have important consequences for the   relative profitabilities of MOET and natural service since MOET allows embryos to be sexed, albeit at higher cost, while natural service (and artificial insemination) does not. Little evidence exists regarding the actual value of F1 or F2 generation male calves, but the value of a male crossbred calf does vary greatly from one region to another. For example, in India, where the value of beef is low and crossbred bullocks are usually insufficiently hardy for draft use, the value of a crossbred male calf is negligible. However, in Latin America, where crossbred calves are used mainly to produce beef, the value of such a calf (at birth) is considerable.

The model presented in Equations 3.1-3.10 was used to analyze the relative profitability of MOET and natural service under varying assumptions regarding the parameters discussed above. For example, MOET should be more profitable the higher the expected F1-F2 differential. The average F1-F2 milk yield differential observed in the experimental trials reported by Cunningham and Syrstad (1987) is 500 kg to 600 kg, although their data evidenced substantial variation in the differential across breeds, countries and within individual trials. Commercial producers also may expect lower differentials than those shown in experimental trials. The model is evaluated using three possible F1-F2 differentials: 400 kg, 500 kg, and 600 kg. F2 cows are assumed to yield 1800 kg per lactation.

F1 cows require higher feed inputs to achieve higher milk yields. The higher feed inputs are calculated using a formulation similar to that utilized by Marion and Wills, 1990, assuming two-thirds the marginal cost and one half the fixed cost for the base situation, consistent with the lower total milk yields of developing country cows, i.e., total feed costs equal $100 + $2.48 per cwt ($0.054 per kg) milk produced. Changes in fixed costs affect the value of MOET only insofar as they affect the profitability of the fifth lactation. Changes in the cost of feed were not considered, though a higher cost of food should increase the relative profitability of MOET.

Two assumptions were used for the price of milk: $0.20 and $0.24 per kg. The lower price is close to the free-trade international price of dry milk during the last decade. The higher price is close to that observed in the majority of developing countries. MOET will be relatively more profitable in areas with a higher milk price.

MOET will be relatively more profitable where its probability of success is higher. Three probabilities were considered for the success rate of an individual application of MOET: .25, .30, .35. In the United States, farmers obtain a successful conception using frozen embryos about 40 percent of the time (Gary Anderson and Eric Bradford, personal communication). Given the much less favorable environment in developing countries, the success rates achieved with embryo transfer are likely to be significantly lower than in the United States, at least until much more experience is gained. American Breeders Service believes that it can achieve a conception rate of 0.35 in developing countries, but such a high rate is surely only possible in areas with a fairly advanced rural infrastructure and then only on the very best dairy farms. The probability of success with natural service was set at .50.

Assuming up to three tries per cow as needed, each with a 40 percent success rate, the use of MOET yields a 78 percent calving rate. Alternatively, the probabilities imply that, on average, 2.5 tries will be required to achieve a successful pregnancy. As the success rate declines, the number of transfers and thus the direct cost per pregnancy rise proportionately. For example, with a 25 percent success rate, each pregnancy is 60 percent more expensive than with a 40 percent calving rate. See Table 3.4. Moreover, lower success rates imply longer calving intervals, postponing the revenue from milk and calf outputs and increasing the probability that a cows's useful life will terminate with fewer total pregnancies and lactations.[22]

TABLE 3.4   MOET Success Rates, Calving Rates, and Number of Embryo Transfers Per Successful Pregnancy

Success Rate	Calving Rate (3 tries)	Tries/ Pregnancy
.25	.58	4.00
.30	.66	3.33
.35	.72	2.86
.40	.78	2.50

Source: Author's calculations.

Finally, MOET and natural service have costs. These costs must be subtracted from the expected value of the F1 or F2 calf produced, respectively, in order to determine the potential profitability of the approach. The costs of natural service are fairly small and are set at $15. The costs of MOET are more difficult to determine.

F1 embryos can be produced by fertilizing embryos from *Bos taurus* cows with semen from *Bos indicus* bulls, or vice-versa. The cost of undertaking MOET in the United States is approximately $250-$300. This includes the cost of flushing the embryo, fertilization, and its transfer to a recipient cow. Of this cost, approximately 25 percent is for drugs, 25 percent for fixed costs such as transportation and refrigeration, and 50 percent for labor. This cost might be reduced in developing countries, since labor is less expensive, but probably not below $200. Costs of this amount are too high relative to the increased productivity achieved through the F1-F2 differential to make MOET profitable.

However, American Breeders Service has developed a method of producing "generic" ovaries at a substantially cheaper price. Using the ovaries from high grade Holstein cows slaughtered in the United States and semen from purebred domestic sires from developing country regions where the embryos will be used, American Breeders Service estimates that it could create *in vitro* fertilized embryos for $35 to $50, and that the cost of transferring this embryo to a local cow would be $15 to $25. Thus, the total cost of creating the fertilized embryo and transferring it to a recipient cow would be $50-$75.[23] The embryos can be sexed for an additional $25-$35. These estimates suggest that it may be possible to transfer sexed F1 embryos to recipient cows in some developing countries for $60-$110. The lower cost seems implausible currently. The figures used here range from $100 to $150 per transfer.

Model Results

The relative profitability of using embryo transfer (generic embryos and embryo transfer) and natural service was calculated using the model set forth above, under different sets of assumptions. The initial assumptions and results are set forth in Table 3.5. The assumptions assume relatively optimistic conditions and success rates, and the results suggest that the profitability of using embryo transfers to create F1 female calves is quite high under such assumptions. The higher milk yield and the higher fertility, in the form of an extra bearing year, lead to a value for an F1 female calf that is roughly twice as high as that for an F2 female calf. Even with a moderate success rate for embryo transfers of 0.3, the dairy producer would achieve a benefit of about $300 per cow per year from adopting a strategy of up to four embryo transfers, relative to a benefit of about $170 from a similar strategy that relied on natural service. The differential of about $130 per cow annually is the expected financial gain from switching from natural service to embryo transfer.

Although the average profitability shown in Table 3.5 is high, the financial risk inherent in using embryo transfers is significant, particularly if it is used on a small number of cows. Because the expected success rate will be low, e.g., 30 percent, on average, dairy producers would have to be prepared to pay for more than three embryo transfers per successful pregnancy achieved. Producers will also have to wait for at least four years before any of the benefits of the F1 female will begin to be realized. The overwhelming number of producers in less developed countries will not have the capital to permit such a strategy, and the risks may be unacceptable even if financing is available.

The effect of varying the probability of success with embryo transfers is shown in Table 3.6. All other parameters are the same. Embryo transfer remains profitable even if the success rate is only 0.25, though the expected gain from using embryo transfer instead of natural service is reduced to about $60 per cow per year.[24]

The effect of varying the cost of embryo transfers is shown in Table 3.7. If the cost of a sexed embryo is assumed to be $150 instead of $100, the expected gain from using embryo transfer is only $12 per cow. Relatively small changes in the cost of embryo transfer have a strong impact on the expected value of using embryo transfer as a breeding strategy since the latter depends importantly on several repetitions, each of which involves a substantial cost.

The effect of not sexing the embryo is shown in Table 3.8. Although the cost of embryo transfer is reduced, the expected value of the expected F1 calf falls sharply and the profitability of embryo transfer is reduced dramatically. Transfer of an unsexed embryo is unprofitable even if the

Table 3.5 Comparative Net Benefits from Use of Artificial Insemination and from Embryo Transfer Under Baseline Assumptions.

F2 female calf, produced with natural service, having productive life of 7 years, 4 bearing years

Assumptions	
Milk yield per lactation	1800 kg
Feed and maintenance, annual costs in first 3 years	$161
annual costs in last 4 years	$258
Price of milk, per kg	$0.24
Cost of artificial insemination	$15
Value of F2 male calf at birth	$100
Probability of natural service success	0.5
Results	
Present value of milk production	$965
Present value of calf production	$401
Present value of feed and maintenance costs	$1025
Value of F2 female calf at birth	$350
Expected present value of natural service	
per try	$90
per four successive tries	$168

F1 female calf, produced with embryo transfer, having productive life of 8 years, 5 bearing years

Assumptions	
Milk yield per lactation	2400 kg
Feed and maintenance, annual costs in first 3 years	$161
annual costs in last 4 years	$290
Price of milk, per kg	$0.24
Cost of transfer of F1 female,	
sexed embryo (no male embryos)	$100
Probability of embryo transfer success	0.3
Results	
Present value of milk production	$1538
Present value of calf production	$479
Present value of feed and maintenance costs	$1226
Value of F2 female calf at birth	$791
Expected present value of embryo transfer	
per try	$121
per four successive tries	$306
Expected financial gain per breeding cow per	
year from use of embryo transfer instead of natural service	$138

Source: Author's calculations.

Table 3.6   Effect of Varying the Probability of Success of Embryo Transfer on the Expected Net Benefit from Embryo Transfer

Assumptions: Same as in Table 3.5, but varying α, the probability of achieving a successful pregnancy with embryo transfer.

Results

α		
0.25	Expected present value of embryo transfer	
	per try	$84
	per four successive tries	$230
0.30	Expected present value of embryo transfer	
	per try	$121
	per four successive tries	$306
0.35	Expected present value of embryo transfer	
	per try	$158
	per four successive tries	$370
	Expected present value of natural service	
	per try	$90
	per four successive tries	$168
	Expected financial gain per breeding cow per year from switch to embryo transfer if probability of success is	
	0.25	*$62*
	0.35	*$202*

Source: Author's calculations.

value of a male F1 calf is raised significantly, e.g., to $250 or more (result not shown). Sexing embryos will be profitable in those areas where the values of male and female calves differ. Moreover, the ability to sex embryos will often be a major attraction of embryo transfer.

The effect of varying the price of milk is shown in Table 3.9. Although a lower milk price reduces the profitability of using embryo transfer to produce an F1 female calf, the use of embryo transfer instead of natural service still achieves an expected financial gain of $60.

The effect of varying the F1-F2 differential between 400 kg and 600 kg is shown in Table 3.10. The gains from embryo transfer are reduced, as expected, if the differential milk yield achieved by the F1 female calf is reduced. However, under the assumptions used in this example, the strategy of using embryo transfers still achieves an expected financial gain of about $68 per cow per year.

The effect of varying the reproductive advantage of the F1 female calf is shown in Table 3.11. The probability of achieving the fifth bearing year is varied between 0.2 and 1.0. In each case, the use of embryo transfer achieves at least a small expected financial gain.

Table 3.7 Effect of Varying the Cost of Embryo Transfer on the Expected Net Benefit from Embryo Transfer

Assumptions: Same as in Table 3.5, but with a varying cost of embryo transfer.

Cost of Embryo Transfer

$100	Expected present value of embryo transfer	
	per try	$121
	per four successive tries	$306
$150	Expected present value of embryo transfer	
	per try	$71
	per four successive tries	$180
	Expected present value of natural service	
	per try	$90
	per four successive tries	$168
	Expected financial gain per breeding cow per year from switch to embryo transfer if cost of embryo transfer rises to $150	$12

Source: Author's calculations.

Table 3.8 Effect of Not Sexing Embryo on the Expected Net Benefit from Embryo Transfer

Assumptions: Same as in Table 3.5, but with no sexing of embryo and associated reduced cost of embryo transfer.

*Embryo type and
cost of embryo transfer*

Sexed, $100	
Expected present value of embryo transfer	
per try	$121
per four successive tries	$306
Not sexed, $75	
Expected present value of embryo transfer	
per try	$49
per four successive tries	$125
Expected present value of natural service	
per try	$90
per four successive tries	$168
Financial loss from switch to embryo transfer if embryo is not sexed	($43)

Source: Author's calculations.

Table 3.9 Effect of Varying the Price of Milk on the Expected Net Benefit
from Embryo Transfer

Assumptions: Same as in Table 3.5, but with a varying price of milk.

Price of milk

$0.20	Expected present value of embryo transfer*	
	per try	$81
	per four successive tries	$206
	Expected present value of natural service	
	per try	$78
	per four successive tries	$146
$0.24	Expected present value of embryo transfer	
	per try	$121
	per four successive tries	$306
	Expected present value of natural service	
	per try	$90
	per four successive tries	$168
	Expected financial gain per breeding cow per year from	
	switch to embryo transfer if milk price is $0.20 per kg	*$60*

---

* Because a lower price of milk is assumed, the value of the cost of feed and
other maintenance inputs is also assumed to be lower, as is the value at birth of
F2 and F1 calves. The expected net benefit of natural service under the same
assumptions is shown for comparison.

Source: Author's calculations.

The use of embryo transfer was calculated to be more profitable than
the use of natural service under each of the scenarios considered above.
However, in each of these scenarios only one parameter was varied. In
Tables 3.12 - 3.16, several parameters in each scenario are varied. Although
several parameters remain at levels that would favor use of MOET, the
simultaneous change of several other parameters to levels that are
unfavorable to MOET is sufficient to reduce the estimated profitability of
embryo transfer below that of natural service. For example, in Table 3.12,
if the F1-F2 milk and reproductive differentials are both lowered, to 400 kg
and to a 50 percent probability of achieving a fifth bearing year, the
profitability of using embryo transfer essentially disappears. If the cost of
embryo transfer is also increased to $150, dairy producers would suffer a
significant loss from using embryo transfer instead of natural service.

Table 3.10 Effect of Varying the F1-F2 Differential Milk Yield on the Expected Net Benefit from Embryo Transfer

Assumptions: Same as in Table 3.5, but with a varying F1-F2 differential milk yield and with feed and maintenance costs varying in proportion.

F1-F2 Milk Yield
Differential

400 kg	Expected present value of embryo transfer	
	per try	$94
	per four successive tries	$238
600 kg	Expected present value of embryo transfer	
	per try	$121
	per four successive tries	$306
	Expected present value of natural service	
	per try	$90
	per four successive tries	$168
	Financial gain per breeding cow per year from switch to embryo transfer if F1-F2 differential is reduced to 400 kg	*$68*

Source: Author's calculations.

Table 3.11 Effect of Varying the F1-F2 Reproductive Differential on the Expected Net Benefit from Embryo Transfer

Assumptions: Same as in Table 3.5, but with varying F1-F2 differential reproductive factors affecting number of calves and lactations during a cow's productive life.

Probability of F2 achieving 5th
calf and lactation

0.2	Expected present value of embryo transfer	
	per try	$77
	per four successive tries	$196
0.5	Expected present value of embryo transfer	
	per try	$94
	per four successive tries	$237
1.0	Expected present value of embryo transfer	
	per try	$121
	per four successive tries	$306
	Expected present value of natural service	
	per try	$90
	per four successive tries	$168
	Expected financial gain per breeding cow per year from switch to embryo transfer if the F1-F2 reproductive differential is reduced to	
	0.2 probability of fifth bearing year	*$28*
	0.5 probability of fifth bearing year	*$69*

Source: Author's calculations.

Table 3.12   Effect of Varying Sets of Parameters on the Expected Net
Benefit from Embryo Transfer (Version A)

Assumptions: Same as in Table 3.5, but with a low F1-F2 milk differential, moderate F1-F2 reproductive differential, and a varying cost of embryo transfer, as noted.

Assumption Set 1)

> F1-F2 Milk Differential = 400
> Probability of achieving 5th calf and lactation = 0.5
> Cost of embryo transfer = 100
> Probability of success per try with embryo transfer = 0.3

> Expected present value of embryo transfer
> > per try $69
> > per four successive tries $174

Assumption Set 2)

> F1-F2 Milk Differential = 400
> Probability of achieving 5th calf and lactation = 0.5
> Cost of embryo transfer = 150
> Probability of success per try with embryo transfer = 0.3

> Expected present value of embryo transfer
> > per try $19
> > per four successive tries $48

Comparison with natural service

> Expected present value of natural service
> > per try $90
> > per four successive tries $168

> *Possible expected financial gain or loss*
> *from switch to embryo transfer:*   *$6 to ($120)*

Source: Author's calculations.

In Table 3.13, the F1-F2 milk differential is set at 500 kg and there is again a 50 percent probability of a fifth bearing year. However, the probability of success with each embryo transfer is increased to 0.35. With this relatively high success rate, the use of embryo transfer remains profitable even if the cost of embryo transfer is increased to $125, but not if the cost rises to $150.

In Table 3.14, the parameters are the same as in Table 3.13, except that the success rate is again set at 0.30. Embryo transfer remains profitable at a cost of $100, but not at $125.

Table 3.13. Effect of Varying Sets of Parameters on the Expected Net Benefit from Embryo Transfer (Version B)

Assumptions: Same as in Table 3.5, but with a low F1-F2 milk differential, a moderate F1-F2 reproductive differential, and a varying cost of embryo transfer, as noted.

Assumption Set 1)

F1-F2 Milk Differential = 500
Probability of achieving 5th calf and lactation = 0.5
Cost of embryo transfer = 100
Probability of success per try with embryo transfer = 0.35

Expected present value of embryo transfer
    per try $112
    per four successive tries $263

Assumption Set 2)

F1-F2 Milk Differential = 500
Probability of achieving 5th calf and lactation = 0.5
Cost of embryo transfer = 125
Probability of success per try with embryo transfer = 0.35

Expected present value of embryo transfer
    per try $87
    per four successive tries $204

Assumption Set 3)

F1-F2 Milk Differential = 500
Probability of achieving 5th calf and lactation = 0.5
Cost of embryo transfer = 150
Probability of success per try with embryo transfer = 0.35

Expected present value of embryo transfer
    per try $62
    per four successive tries $145

Comparison with natural service

Expected present value of natural service
    per try $90
    per four successive tries $168

Possible expected financial gain or loss from switch to embryo transfer: *$95 to ($23)*

Source: Author's calculations.

Table 3.14   Effect of Varying Sets of Parameters on the Expected Net
Benefit from Embryo Transfer (Version C)

---

*Assumptions: Same as in Table 3.5, but with moderate F1-F2 milk and reproductive
differentials, a moderate rate of embryo transfer success, and a varying cost of embryo
transfer, as noted.*

Assumption Set 1)

> F1-F2 Milk Differential = 500
> Probability of achieving 5th calf and lactation = 0.5
> Cost of embryo transfer = 100
> Probability of success per try with embryo transfer = 0.30

> Expected present value of embryo transfer
> > per try $82
> > per four successive tries $207

Assumption Set 2)

> F1-F2 Milk Differential = 500
> Probability of achieving 5th calf and lactation = 0.5
> Cost of embryo transfer = 125
> Probability of success per try with embryo transfer = 0.30

> Expected present value of embryo transfer
> > per try $57
> > per four successive tries $143

Comparison with natural service

> Expected present value of natural service
> > per try $90
> > per four successive tries $168

*Possible expected financial gain or loss from switch to embryo
transfer:*   *$39 to ($25)*

Source: Author's calculations.

The effect of varying the probability of success with embryo transfer
when several other parameters are set at somewhat less favorable levels is
examined in Table 3.15. The F1-F2 milk and productivity differentials are
set at 500 kg and a 20 percent probability of a fifth bearing year. The cost of
transferring a sexed embryo is $100. The use of embryo transfer remains
profitable if the success rate is high, but not if it is moderate or low.

In Table 3.16, embryo transfer remains profitable with only a moderate
milk differential and a low fertility differential as long as the probability of

Table 3.15   Effect of Varying Sets of Parameters on the Expected Net
Benefit from Embryo Transfer (Version D)

Assumptions: Same as in Table 3.5, but with a moderate F1-F2 milk differential, a low
F1-F2 reproductive differential, and a varying success rate of embryo transfer, as noted.

Assumption Set 1)

F1-F2 Milk Differential = 500
Probability of achieving 5th calf and lactation = 0.2
Cost of embryo transfer = 100
Probability of success per try with embryo transfer = 0.30

Expected present value of embryo transfer	
per try	$66
per four successive tries	$167

Assumption Set 2)

F1-F2 Milk Differential = 500
Probability of achieving 5th calf and lactation = 0.2
Cost of embryo transfer = 100
Probability of success per try with embryo transfer = 0.35

Expected present value of embryo transfer	
per try	$82
per four successive tries	$207

Comparison with natural service

Expected present value of natural service	
per try	$90
per four successive tries	$168

Possible expected financial gain or loss from switch to *embryo transfer:*	$39 to ($1)

Source: Author's calculations.

transfer success is high. However, if the latter is reduced as well, embryo
transfer is not profitable.

The composite results shown in Tables 3.5-3.16 suggest that the use of
embryo transfer to achieve an F1 female calf has potential to be profitable
in areas with relatively favorable conditions. It remains to be seen whether
the sexed embryos can be successfully transferred at the prices which have
been estimated, and whether the F1-F2 differentials are sufficiently large
in practice to make the process profitable. If rather small shortfalls occur
simultaneously in a number of the parameters assumed, embryo transfer
will not be profitable. Conditions in most developing countries are difficult

Table 3.16   Effect of Varying Sets of Parameters on the Expected Net
Benefit from Embryo Transfer (Version E)

---

*Assumptions: Same as in Table 3.5, but with a moderate F1-F2 milk differential, a low
F1-F2 reproductive differential, a high success rate for embryo transfer, and a varying cost
of embryo transfer, as noted.*

Assumption Set 1)

   F1-F2 Milk Differential = 500
   Probability of achieving 5th calf and lactation = 0.2
   Cost of embryo transfer = 100
   Probability of success per try with embryo transfer = 0.35

   Expected present value of embryo transfer
      per try $94
      per four successive tries $220

Assumption Set 2)

   F1-F2 Milk Differential = 500
   Probability of achieving 5th calf and lactation = 0.2
   Cost of embryo transfer = 125
   Probability of success per try with embryo transfer = 0.35

   Expected present value of embryo transfer
      per try $69
      per four successive tries $161

Comparison with natural service

   Expected present value of natural service
      per try $90
      per four successive tries $168

   *Possible expected financial gain or loss from switch
   to embryo transfer:* $52 to ($7)

Source: Author's calculations.

and will make it hard to achieve the performance levels that have been
assumed.

Implementing an embryo transfer delivery system in an developing
country may also imply greater costs than has been estimated. The costs in
developed countries are based on a sophisticated level of communications
and transportation, an ample supply of well-trained technicians, and the
existence of well-educated producers with considerable management skills.
Few developing countries have these advantages. Developing a system to

carry out embryo transfers will require either that the developing countries make substantial improvements in each of these dimensions, or that the centers established to undertake embryo transfers establish logistical and training support to substitute for the lack of these factors.

To establish embryo transfer centers, veterinarians and technicians would have to be trained, communication facilities would have to be established (radios, cellular telephones, bicycles, trucks) to facilitate transport and information flows between the center and dairy producers, and dairy producers would have to be trained in the purposes and functioning of embryo transfers, and they would have to adhere to rather precise schedules.

To save on logistics, it will be important to synchronize the estrus of a significant number of cows in the same herd so that embryo transfer can be carried out simultaneously, with significant savings in technicians' time. Transfers for individual cattle will not be profitable in most cases. The use of prostaglandins to achieve synchronization is but one of the technologies that would be required to bring the costs of embryo transfer to a reasonable level. While the technology is relatively simple, it would require a level of understanding, commitment to detail and to precise timing which is rarely available for developing country dairy herds where cattle are managed by farmers and laborers whose adherence to schedule is often weak and whose scientific understanding is limited. Weak herd management and difficult logistics contribute to the relatively poor performance of artificial insemination in developing countries. For example, Eddebbarh (1987) notes that it proved impossible in Morocco to implement a progeny testing program because it was not understood "by the majority of farmers and, in some situations, not even by technicians." The management and logistical requirements of embryo transfer will be greater yet.

Ironically, while small-scale farmers supervise their animals more intensively, they are less likely to have the technical and commercial sophistication required to interact with a MOET center, and, having fewer animals, could not achieve the economies of scale required to reduce costs to the levels considered in the calculations above. Small-scale farmers also are at a greater disadvantage in terms of access to telephone and transport. Frequently, no phone is available or it is not functioning. It may be difficult, costly and slow for the farmer to travel to the veterinarian's office. The logistical and technical assistance inputs required to work with many small-scale farmers, particularly given their relatively lower educational background and familiarity with sophisticated business transactions, would increase costs still more.

Since many producers would not have credit to pay for relatively expensive embryo transfers for all of their cows each year, an embryo transfer center would probably have to finance the investments (This might be linked through a producers' cooperative). These innovations are feasible over the longer run, but it is unrealistic to expect their implementation in the near to intermediate future.

The available experimental data also indicate that there is a great deal of variation in the performance of F1 cows relative to the parent population, higher number generation cows such as F2 and F3 generations, or even backcrosses (Cunningham and Syrstad, 1987; Turton, 1985). This variation in productivity suggests that even if there is a strong statistical relationship within the larger data set suggesting the relative attractiveness of F1 animals, there is no guarantee that any individual F1 cow will perform well. Further, the optimal proportion of *Bos taurus* genes in any developing country dairy population will depend on the particular production environment (Vaccaro, 1994). Although the F1 cross appears to offer advantages in many regions today, the optimal proportion in other regions may be higher and a different proportion of exotic genes, perhaps produced through rotational or composite breeding, could provide even better results than the F1 generation crossbreed.

Summary of Potential for Use of MOET in Developing Countries

Although many obstacles must be overcome, the examples provided suggest an intriguing potential for the use of embryo transfer to produce F1 generation female dairy calves in many of the world's tropical regions. This approach is not likely to be profitable for the great majority of producers in developing countries in the foreseeable future, but it seems likely that it will contribute in some degree to increasing milk production in developing countries during the next two decades.[25]

MOET will also be used to as a technology to increase the rate of genetic improvement as part of other breeding technologies. Indeed, there is greater opportunity for the long term use of MOET in these programs, probably within the context of an ONBS, than in the production of F1 calves.

Notes

1. In genetics, heterosis is the increase in productivity of the hybrid over that of the parents. Cunningham (1987) defines heterosis in milk production as the difference between the yield of the first generation hybrid cross and the mean of the parents' yields.

2. The use of embryo transfers offers potential to refine the ultimate selection process of females in addition to multiplying superior female germplasm. If several cows are superovulated, evaluation of their calves will theoretically allow dropping some of the donor cows on the basis of calf performance. However, this effect is relatively small given the (currently) limited number of ova that can be obtained per cow.

3. The use of embryo transfers may also play a role in the conservation of indigenous tropical breeds whose existence is threatened as other breeds and/or crossbreeds are substituted for them.

4. The difficulty of field testing in developing countries has made it difficult to evaluate progeny from bulls developed locally, and the lack of information on local bulls has led to greater reliance on the use of imported semen. However, animal performance is affected by a fairly strong genotype-environment interaction, where environment includes climate, disease and parasite pressure, nutrition and herd management (Buvendran and Petersen, 1980). Thus, bulls whose offspring are favorably ranked (relative to the offspring of other bulls) in developed country environments often do not produce offspring which are favorably ranked in less developed country environments (Blake et al., 1989).

5. A cooperative breeding scheme like an ONBS is likely to also achieve broader benefits through improvements in generalized herd management. The effectiveness of extension services has been shown to improve in such schemes (personal communication, Ed Rege).

6. Ned Rege indicates that in some cases the (understandable) reluctance of farmers to sell their best cows to the nucleus unit may be overcome by "borrowing" these animals on a contract. Effectiveness of the ONBS approach may also require the periodic exchange of bulls between herds (Kasonta and Nitter, 1990).

7. In most developing countries, cows are hand-milked. In many areas where *Bos indicus* cattle are used, the calf is kept nearby to stimulate milk letdown. The calf is generally allowed at least one quarter of the milk.

8. However, Vaccaro (1994) emphasizes that significant gains can be had from selection among tropical cattle herds due to the extreme deviations of the performance characteristics of individual cows. Thus, the long run gains from selection among indigenous breeds may be larger than the long run gains from crossbreeding, though crossbreeding can often achieve more rapid results. Ed Rege (personal communication) also notes that indiscriminate crossbreeding threaten the extinction of indigenous breeds in many regions of the world.

Selection as opposed to crossbreeding also has greater scope to allow for genotype variation. A standard genotype has disadvantages insofar as it cannot be tailored to the distinct needs of farms which, although located in the same region, provide different environmental and managerial conditions for their livestock.

9. An F1 animal can be created by crossing local and exotic cattle, using either natural service or artificial insemination. However, if two F1 generation animals

are crossed via natural service or artificial insemination, their calf will be an F2, having an equal mix of local and exotic genes.

10. When fit into a production system framework, the change in milk yields has a much more important effect on total production than does the change in the other two characteristics.

11. The parent breeds of the crossbred animals vary from study to study, often with important impact on base milk yield. The F1-F2 differential observed may be related to the choice of the parents as well as the environment, nutrition, and management received. In the most extreme case, Brazil, data were reported for the milk yield differential between generation 1 and generation 2 Pigangueiras cattle, a composite breed carrying about 3/8 *Bos indicus* (Buzerat) and 5/8 *Bos taurus* (Red Poll) inheritance. The differential in milk yield in two experiments was about 14 percent.

12. The estimated differentials again show considerable country to country variation. The latter variation, as well as its causes, has important economic implications for determining whether and where the use of MOET for continuous production of F1 generation calves might be profitable.

13. The F1-F2 differentials reported above could exaggerate the increase in milk yields that commercial farmers would achieve by using F1 generation instead of F2 (or other) generation cows. Management probably affects the absolute level of yields and would affect the F1-F2 differential if the latter is proportional to F2 output levels. Management was probably of higher quality on the experiment stations where the performance of different crosses were studied than it will be on commercial farms. There is some evidence that new technologies, when adopted by farmers, achieve about two-thirds of the yields achieved under experimental farm conditions (Davidson and Martin, 1965). If the F1-F2 differential discovered on experimental farms behaves similarly, the expected differential on commercial dairies in India would be about 375 kg instead of about 570 kg.

14. Although the production of F1 offspring is usually the first step in any exotic crossing program in a *Bos Indicus* population, the "continued use of F1 animals beyond the first generation is operationally difficult and the relative sizes of the additive and heterosis effects has not been economic given its high costs" (Cunningham and Syrstad, 1987).

15. For simplicity, it is assumed here that the alternative to MOET is natural service.

16. The probabilities of success of each try are considered independent.

17. This is true if the probabilities of success on successive tries are statistically independent. In practice, if attempted once, MOET is usually repeated several times. If MOET fails to produce a calf after three tries, a farmer may try natural service or artificial insemination. That possibility is ignored here; the impact on the calculated values is small because the event occurs with low probability.

18. It is possible that:
$(\alpha X-B)(1+(1-\alpha)+(1-\alpha)^2+....)=(\alpha X-B)(1/1-\alpha) > (\beta Y-A)(1+(1-\beta)+(1-\beta)^2+....)=$
$(\beta Y-A)(1/1-\beta)$, even if $(\alpha X-B) < (\beta Y-A)$, and vice-versa.

Thus, a salesperson for MOET might correctly tell a farmer that although the present value of each single try at natural service would be higher than the present value of each single try at MOET, the farmer would be better off to adopt a strategy of several repetitions of MOET instead of several repetitions of natural service.

19. It is difficult to calculate their values simultaneously since the value of a female calf depends importantly on the value of its future offspring, leading to potentially explosive interactions among parameters. In the calculations, a plausible present value for a female calf was selected and, given this value for female calves and a plausible value for a male calf, an average expected calf value was obtained. This average expected calf value was then used in Equations 5 and 6, and feed costs were manipulated to equate the values of F1 and F2 female calves with the values that had been assumed.

20. Theoretically, these equations should also contain the discounted value of meat production at slaughter. However, the difference in these terms should be small. Lacking specific information to the contrary, for simplicity, the terms are considered offsetting in the two equations and are omitted.

21. If the use of MOET to produce F1 calves is profitable, all future calves (produced out of the initial F1 calves) should be produced by MOET as well. However, if F1 and F2 generation cows are assumed equally capable of serving as donors, the use of MOET confers no differential value on the calves of F1 and F2 cows. Each would then bear F1 calves in the future. If that is true, the benefits from using MOET to produce the initial F1 calf depend only on the F1-F2 milk differential. If the F1-F3 differential is lower, it might not be profitable to use MOET on F2 generation cows even if it were profitable to use MOET on F1 generation cows.

22. Because the magnitude of the fertility differential is not precisely known, the two terms corresponding to the eighth year in Equation 3.7 are multiplied by a parameter representing the probability of obtaining a fifth calf. This parameter can be varied to observe its effect. For example, if F1 cows regularly achieve 5 calves (and lactations), with the last occurring in year 8, this probability is 1. If, instead, only half the F1 cows are expected to achieve 5 calves and lactations, the probability is 0.5. For simplicity of notation, this parameter is not shown explicitly in Equations 3.5 and 3.6).

23. The costs cited do not include the costs associated with changes in management or other farm level inputs that would be needed for successful implementation of MOET on a developing country dairy farm. These costs would likely be substantial.

24. With a success rate of 0.25, the profitability of one try with embryo transfer is lower than that of one try with natural service, though the relative profitabilities reverse if the farmer commits to four tries.

25. Other benefits from the use of embryo transfer might increase the value of continuous F1 production. For example, assuming no problems with dystocia, male embryos of a different breed could be chosen to increase beef production.

4

The Projected Effects of rbST and MOET on World Dairy Markets

Lovell Jarvis and Atanu Saha

This study began with a concern regarding the effect that two new biotechnologies, rbST and MOET, might have on world dairy markets. The chapters on the expected profitability of rbST and MOET concluded that each technology is likely to be profitable immediately in the developed countries and therefore likely to have progressively greater effect over time on milk output in such countries. It seems unlikely that either rbST or MOET will be profitable in developing countries in the short run, except in exceptional circumstances. However, in the longer run both rbST and MOET should be widely adopted throughout the world.

This chapter analyzes the longer run implications of these findings for world dairy markets. It first discusses the probable effects of rbST and of MOET on dairy production in the developed and developing countries. Second, it presents a broad overview of recent trends in the production, consumption and international trade in milk. Third, it projects trends, under a range of plausible scenarios, to obtain a sense of the supply-demand balances likely to hold in the years 2000 and 2010 under different assumptions regarding the degree of adoption of rbST and MOET in both developed and developing countries. Finally, it considers the implications of the projected supply-demand balances.

The Effects of rbST and MOET Adoption on International Milk Markets.

The effects of rbST and MOET on developed country comparative advantage in milk production will depend on the degree to which unit costs are reduced in developed countries relative to those in developing countries. This difference will depend significantly on the factors determining the production response to rbST and to MOET in both areas and also on how the use of rbST and MOET interacts with the development of other technologies to further increase dairy efficiency.

rbST is likely to improve the international competitive position of developed country milk producers both in the short and long-run. rbST adoption should occur more rapidly in developed countries than in less developed countries, and it should cause a larger reduction in unit costs once adoption has reached its limit. MOET is also likely to be adopted more rapidly in developed countries, though there is less reason to expect that its adoption will reduce unit costs more in developed than in developing countries. Indeed, the contrary could be true.

For example, if developed country milk markets are increasingly liberalized as a result of milk output increases stemming from rbST adoption, the downward pressures on the price of milk could cause relatively more efficient producers to displace the relatively less efficient producers more rapidly than otherwise might occur. In the United States, as noted, farms having larger numbers of higher yielding cows would seem to have greater incentive to adopt rbST. Because such farms generally already have lower unit costs than farms having smaller, lower yielding cow herds, rbST adoption could promote a structural shift in the U.S. milk industry that would cause U.S. milk productivity to rise fairly sharply. Such structural changes have been slowed by the price support systems which have long characterized the U.S. market.[2]

The problem for the world milk market and especially for less developed country milk producers is that international milk prices can be affected in the short-run in substantially greater degree than the reduction in long-run costs. If rbST adoption provokes a short-run price decline in developed countries sufficiently large so as to cause severe hardship for non-adopting farms, developed country governments could attempt to forestall the high social costs of such adjustment by imposing milk production quotas or by accumulating milk stocks, and engage in forceful international dumping as occurred in the 1980s (Larsen and Brown, 1987). The United States, which faced significant dairy surpluses in the 1980s, moved to reduce supply through a herd reduction program and a more flexible price support program. Due to a change in U.S. farm legislation, the U.S. support price fell from about $13.20 per cwt in 1981 to $10.00 per cwt in 1990, and dairy stocks also fell. The Economic Community also moved during the mid-1980s to reduce and control supply, mainly through the use of production quotas, and has substantially reduced surplus production and milk stocks.

Passage of the Uruguay Round makes it seem increasingly likely that developed country governments will either allow domestic prices to decline as output increases, thus allowing the market to equilibrate, or will utilize milk quotas to control output as unit costs of production decline. Milk prices declined somewhat during the first months after the use of rbST was approved, but it is still to early to determine cause and effect. If prices decline significantly, the government will come under pressure to restrict output.

Milk quotas could be used to restrict output, allowing adopters of rbST to reduce costs and increase profits without causing a severe decline in milk price. If quotas are allowed to be marketed, the cost savings from use of rbST will quickly be capitalized into quota values, encouraging less efficient firms to sell quotas to more efficient firms while allowing the less efficient firms significant capital protection. However, consumers will reap no gains from the cost savings and, since many consumers apparently remain concerned about the human health risks posed by rbST, consumer opposition to rbST and to quotas might intensify.[3]

Dairy production in developing countries rose rapidly in the 1970s and 1980s (see below), significantly more so than in the developed countries. Horizontal expansion, conventional management improvements, and the availability of technologies other than these new biotechnologies provide substantial scope for continued output expansion. The adoption of rbST in developed countries is not likely to greatly slow the rise of output in the developing countries, even if it causes international prices to decline somewhat. Most developing countries protect their dairy sectors so that their producers are not fully exposed to the movements in the international milk price, although the incentive to export is affected. Because many developing countries currently are net importers of milk products, their total welfare could be improved by a lower international milk price. However, producers and consumers will see this issue differently, and some developing countries could become important exporters in the future.

The World Dairy Industry: A Glance Backwards

This section presents a broad overview of recent trends in the production, consumption and international trade of milk, and projects trends, under a range of plausible scenarios, to obtain a sense of the supply/demand balances likely to hold in the years 2000 and 2010. These scenarios include different assumptions regarding the degree of adoption of rbST and MOET in the dairy industry in both developed and developing countries. From these projections, insights may be obtained regarding the probable trend in milk prices, as well as in the international pattern of production, consumption, and trade under different assumptions regarding the adoption of these two biotechnologies.

The usual note of caution that is associated with the use of simple projections is in order here. The projections are utilized as a simple and thus rather quick means to assess future supply/demand balances. Projections implicitly assume that the set of conditions which governed past developments are expected to persist over the relevant period. However, the scenarios being considered include cases where partial diffusion of the two biotechnologies occur. It would be surprising if diffusion occurred without producing any feedback. For example, if adoption led to an increase in milk production, the price of milk should fall, and milk consumption should rise more rapidly. Because the projections do not explicitly model such interactions, the projections are only a rough indicator of potential market changes.

In most, although not all countries, the supply of milk demonstrates fairly high short- and long-run price elasticities. The magnitudes of these elasticities are reflected in the significant changes in the rates of growth of milk production that have been observed for individual countries over fairly short periods. That rates of growth can change significantly within a short period makes it difficult to project trends with confidence. In addition, governments have frequently intervened in milk markets, deeming the free-market price politically unacceptable. Recent pronouncements by governments and international agreements suggest that developed country governments that have intervened forcefully in the past have decided to intervene less. If so, a major market determinant in the immediate past may have changed, requiring further adjustment when projections are made. Although these factors might be treated better in an econometric model, any such model would have to be large, complex and costly. An accurate explanation/prediction of government behavior might well remain elusive. The projections presented here are offered as a rough substitute. Several scenarios are presented in order to analyze the sensitivity of the results to changes in the assumptions.

This overview covers 117 countries of which 91 are classified as developing and 26 as developed. Of the developing countries, 12 are in Asia, 51 in Africa and 28 in Latin America. Countries are further subdivided into geographical subregions. The countries and their classifications are presented in Appendix I.

Past trends are calculated from data for the period 1961-1988. Several projections of output, consumption and the resulting surplus or deficit in milk in the years 2000 and 2010 are presented, each depending on a slightly different scenario. These scenarios differ primarily in their assumptions regarding the projected growth rates of milk production, which may be affected by the adoption of rbST and/or MOET, the growth rate of per capita income, and the income elasticities which link income growth to milk consumption in individual countries. The methodology adopted is set out in Appendix II.

The data on production, consumption and trade used in this study are from the Food and Agriculture Organization of the United Nations (FAO) and the population statistics and economic projections are from the World Bank. Milk production and consumption are defined to include fresh and dry cow, sheep, goat, and buffalo milk. Although cow milk accounts for the bulk of milk in most countries, other types of milk are important in parts of Africa and Asia, especially in India. The inclusion of sheep, goat, and buffalo milk increases significantly the share of milk that is produced in developing countries.

Production

World milk production was 494 million tons in 1988 (see Table 4.1). The developed countries produced 74 percent of total milk output, although these countries contained less than a quarter of the world's population.

Table 4.1 Production of Milk in 1988

	Production (million m. tons)
Asia	72.3
Africa and Middle East	25.9
Latin America	39.3
EEC & Western Europe	129.5
Eastern Europe & USSR	133.2
US & Canada	72.7
Australia & N. Zealand	13.6
Developed Countries	356.3
Developing Countries	137.5
World	493.8

Source: Authors' calculations using data from FAO Agricultural Production Tapes.

Per capita milk production differs widely among regions and among countries, even among developed countries. Per capita milk output in the developed countries averaged nearly 300 kgs annually, about 9 times that of the developing countries. Regionally, the highest per capita output, about 670 kgs, is found in Oceania and the lowest, about 40 kgs, is found in Asia and Africa. The observed difference in average per capita production between the developed and developing countries is determined primarily by the difference in average per capita consumption, reflecting the large difference in average per capita income in the two regions. However, per capita production and consumption also vary across countries having similar income levels, revealing the significant influence of culture (tastes), natural endowments and accumulated investments.

Output has grown consistently more rapidly in the developing

Table 4.2   Spatial Distribution of Milk Production Over Time

(percentage of world production)

	1961-70	*1971-79*	*1980-88*
Asia	9.8	9.5	12.0
S. Asia*	2.1	1.9	2.0
E. & S.E. Asia**	0.2	0.2	0.5
India	6.4	6.1	8.1
Africa and Middle East	6.2	5.6	6.1
N. Africa	1.1	1.1	1.5
Middle east	2.7	2.4	2.5
W. Africa	0.4	0.3	0.3
C. Africa	0.1	0.1	0.1
E. & S. Africa	1.9	1.7	1.7
Latin America	7.1	7.3	7.6
C. Am. & Ca.	1.7	2.2	2.2
S. America	5.4	5.1	5.4
Japan	1.1	1.3	1.4
EEC	10.3	24.8	24.1
W. Europe	8.7	4.5	4.1
E. Europe	10.7	9.8	9.4
USSR	23.1	21.1	19.7
US & Canada	20.2	14.5	14.4
Austr. & N.Z.	4.1	3.0	2.7
Developed	77.0	77.7	74.4
Developing	23.0	22.3	25.6

*excluding India.
**excluding Japan

Source: Authors' calculations using data from FAO Agricultural Production Tapes.

countries than in the developed countries during the last three decades. Given the small initial share of the developing countries, the rather small absolute differences in growth rates have not yet significantly changed output shares. This will soon change, however, if current rates of growth persist (Table 4.2).

Past growth rates of milk production for individual countries were estimated by regression analysis using the relationship:

$Q_t = Q_0(1+r)^t$, where

Q_0 = output in the first year of the period,

Q_t = output in the last year of the period, and

r = the trend rate of growth.[4]

Most of the estimated trend coefficients were statistically significant, although the growth rates obtained from these regressions usually differed from those which could be calculated using period end points. For example, although the developing countries' share of world milk production declined slightly in 1971-79, before rising in 1980-88 (Table 4.2), the estimated trend growth rates are higher for the developing countries than for the developed countries in each period.

Somewhat surprisingly, the growth rate of world milk output steadily declined during the last three decades (Table 4.3). This is true for both developed countries and developing countries, taken as country groups. However, among developing country regions, Latin America and Africa each experienced a steady decline in the growth rate of milk output, while Asia experienced a steady rise. See Table 4.3. The slight rise in the share of the developing countries in the 1980s stems from an acceleration of milk production growth rates in several Asian developing countries, and a deceleration in production growth rates in the developed countries.

A few countries account for most of the milk produced in developing countries. For example, in 1988, India produced 46 million tons of milk, more than 67 percent of Asia's total output, more than the total output of either Latin America or Africa, and roughly three times the combined output of Australia and New Zealand. Pakistan was also an important producer with 13 million tons. Brazil, Mexico and Argentina were the major producers in Latin America with outputs of 14, 8, and 6 million tons respectively. Turkey was the largest producer in the Middle East with 5 million tons, and Sudan was the largest producer in North Africa with 3 million tons.

Although several other developing countries produce significant amounts of milk, the seven countries mentioned above produced 75 percent of total developing country milk output in 1988. They comprised, however, only 51 percent of developing country population (excluding China).[5] Production in these seven key developing country milk producers thus averaged about three times as much on a per capita basis as did production in other developing countries.

The principal change during the past three decades in the spatial distribution of milk output within the developed countries has been the

Table 4.3   Annual Growth Rates of Milk Production

	1961-70	1971-79	1980-88	1961-88
Developing	2.71	2.58	2.11	2.68
Asia	2.61	2.82	3.80	3.45
S. Asia*	2.27	2.62	1.04	1.98
E. & S.E. Asia**	6.18	4.15	3.48	4.60
India	0.01	3.04	4.89	2.65
Africa & Middle East	2.11	2.20	1.85	2.08
N. Africa	1.93	4.40	3.04	3.12
Middle East	2.77	2.86	2.02	2.55
W. Africa	3.38	0.74	1.28	1.80
C. Africa	2.76	2.07	1.65	2.16
E. & S. Africa	1.20	1.78	1.23	1.41
Latin America	2.99	2.37	1.89	2.39
C. Am. & Ca.	2.96	2.14	0.42	1.84
S. America	3.03	2.43	2.11	2.58
Developed	1.35	1.23	0.89	1.15
Japan	8.44	2.34	1.77	4.18
EEC	1.18	2.02	0.52	1.24
W. Europe	0.92	1.68	-0.34	0.75
E. Europe	1.37	3.27	1.02	1.89
USSR	2.91	1.57	1.49	1.99
US & Canada	0.32	0.13	0.90	0.15
Austr. & N.Z.	2.03	-1.01	1.66	1.15
World	2.28	1.83	1.50	1.87

* excluding India
**excluding Japan

Source: Authors' calculations using data from FAO Agricultural Production Tapes.

substantial decline in the share of the United States and Canada, and the rise in the share of Europe (European Community-12 and Western Europe). Europe's share increased from 19 percent in the 1960s to 29 percent in the 1970s, while the United States and Canada's share declined from 20 percent to 15 percent.

Among developed and developing countries, the performance of individual countries and, even regions, was highly variable over the last three decades. See Table 4.4. For example, although milk production in the developed countries grew during the 1980s at less than one percent, with the rate being particularly low in Europe, the United States and

Table 4.4   Growth Rates of Milk Production for 1961-1988, by Region
(in percentages)

	0 - 1%	1 - 2%	> 2%
DCs	W.Europe US and Canada	EEC E.Europe USSR Austr. and N.Z.	Japan
LDCs		S.Asia W. Africa E.& S. Africa C. Amer. & Carib.	E. & S.E. Asia India C. Africa N. Africa Middle East S. America

Source: Authors' calculations using data from FAO Agricultural Production Tapes.

Table 4.5   Growth Rates of Milk Production for 1980-1988, by Region
(in percentages)

	< 0	0 - 1.5%	1.5 - 2%	> 2%
DCs	W. Europe	EEC E.Europe USSR US and Canada	Japan Austr. and N.Z.	
LDCs		S.Asia W. Africa E.& S. Africa C. Amer. & Carib.	C. Africa Middle East	E. & S.E. Asia India N. Africa S. America

Source: Authors' calculations using data from FAO Agricultural Production Tapes.

Canada, milk production grew at a significantly higher rate in Japan, Oceania and the USSR. See Table 4.5. Growth rates in developing countries were similarly diverse. While milk production growth rates in East and South East Asia, India, and North Africa were high, rates of growth in South Asia, West Africa, and Central America and the Caribbean were low.

In 1980-88, the growth of milk output in Asia accelerated to 3.8 percent, a rate nearly double that observed in Latin America or Africa. Several

Table 4.6 Annual Growth Rates of Milk Output (Selected Asian Countries)

	1961-70	*1971-79*	*1980-88*	*1988 share of Asian Production*
India	0.01	3.04	4.89	67.20
Pakistan	2.1	1.6	3.4	19.51
Korea Rep.	33.9	20.3	32.1	2.04
Indonesia	2.8	11.5	14.7	0.33

Source: Authors' calculations using data from FAO Agricultural Production Tapes.

Asian countries increased production sharply, as shown in Table 4.6. For example, although it began the period with a low base, South Korea sustained a growth rate of milk production exceeding 25 percent per annum for three decades. India, which produced substantial milk at the beginning of the period, steadily increased its rate of growth of milk output to nearly 5 percent per annum in the 1980s.[6]

In Latin America, the growth rate of milk output was considerably lower in the 1980s than in the previous two decades (although a recovery began at the end of the decade and carried into the 1990s). See Table 4.7. Milk growth rates in Latin American countries were highly variable during the last decade. The decline in milk production was mainly a result of declining local demand for milk in response to falling consumer incomes. The decline in demand is also reflected in a declining growth rate of food production in this region. The growth rate of food output was 4.2 percent in 1961-70, 1.8 percent in 1971-80, and 1.3 percent in 1981-88 (FAO data; not shown).

Milk production has been increasing at a very low rate in Sub-Saharan Africa (comprising the sub-regions West, Central, East and South Africa).

Table 4.7 Annual Growth Rates of Milk Output (Selected Latin American Countries)

	1961-70	*1971-79*	*1980-88*	*1988 Share of Latin America Production*
Mexico	4.42	6.65	0.37	19.9
Argentina	0.96	1.71	1.96	16.8
Brazil	3.91	4.49	0.92	34.4

Source: Authors' calculations using data from FAO Agricultural Production Tapes.

Although there were some success stories in this region, such as the growth achieved via Kenya's Integrated Dairy Project, overall progress has been slow.

The slow growth of milk output in the United States, Canada, and the European Community is explained mainly by sluggish demand in these regions. A system of price supports and programs for the disposal of surplus production has kept domestic prices at high levels, avoiding the significant production decline that would otherwise have occurred. A high level of accumulated stocks induced the European Community to subsidize milk exports, especially during the early 1980s, resulting in depressed international milk prices.

Although production has been rising, milk yields in developing countries remain low relative to those in developed countries. Moreover, with the exception of a few Asian countries where rapidly rising yields were a major contributor to rising milk production, milk yields rose slowly or not at all in most developing countries. This result is in contrast to the experience of developed countries, where milk yields rose steadily during the last three decades and is the major factor explaining the increase in milk production.[7] In most developing countries, it was more profitable to expand the number of dairy herds than to increase output per cow, given the prevailing price ratios and the resources and technology available. This situation is beginning to change as the rising value of land creates pressures to increase yields.[8] There is great scope for increasing milk yields in developing countries through improvements in animal nutrition, farm management, veterinary care, and, in the long run, the genetic composition of many herds.

Market intervention in the form of price supports to producers is a feature not peculiar to the developed countries. In the face of declining international milk prices and important milk import deficits, many developing countries protected their domestic dairy industry via import tariffs and price supports. Such protection became prominent during the early 1980s, when international prices fell dramatically, and the rapid growth of milk production in certain developing countries was partially explained by that protection. Data on the degree of protection in the early 1980s for selected developing countries is given in Table 4.8. The "nominal protection" coefficients are measured as the ratio of domestic price to world market price at the respective country's border, where world market price is defined as the New Zealand export price plus transportation cost from New Zealand to the country of concern. The level of protection in these developing countries was quite high but declined as international prices recovered during the late 1980s.

Table 4.8   Self-Sufficiency Ratios and Nominal Protection Coefficients

	1980-82, Self Sufficiency Ratio (%) [a]	1980-82, Nominal Protection Coefficient [b]
DCs		
Australia	124	1.30
Canada	111	1.95
EC -10 [c]	113	1.75
EFTA -5	111	2.40
Japan	84	2.90
New Zealand	219	1.00
Spain and Portugal	95	1.80
United States	103	2.00
USSR	98	2.60
East Europe	101	2.60
LDCs		
Egypt	75	2.50
Nigeria	32	3.00
South Africa	98	2.30
China	95	2.80
India	99	1.80
Korea	98	3.00
Taiwan	93	3.00
Thailand	6	1.80
Argentina	100	1.00
Brazil	99	1.60
Mexico	90	2.80

a=ratio of local production to local consumption times 100,
b=ratio of local producer price to border price,
c= European Free Trade Association, including Austria, Iceland, Norway, Sweden, and Switzerland.

Source: Hallberg and Cho, 1987.

Consumption

Like production, the consumption of milk is concentrated in developed economies. In 1988, the developed countries accounted for 68 percent of the total consumption of milk, with a per capita consumption of 303 kg. The corresponding figures for the developing countries were 32 percent and 49 kg. See Table 4.9.[9] There is also considerable variation in the level of

Table 4.9   Per Capita Consumption of Milk in 1988, by Region

	(kgs)
Asia*	42.8
Africa & Middle East	39.4
L. America	91.5
EEC & W. Europe	392.7
E. Europe	330.1
US & Canada	264.1
Austr. & N. Zealand	676.9
DCs	303.4
LDCs	49.1
World	113.6

*Excluding China. In 1988, the per capita consumption in China was 2.9 kg.

Source: Authors' calculations, using data from FAO Agricultural Production Tapes.

milk consumption among developing country areas, with that in Asia and Africa and the Middle East being only about half that of Latin America. The variation among individual countries was much greater still, from about 1 kg annually in Indonesia to about 300 kg in Uruguay.

Milk consumption has grown in all regions when measured over the last three decades, although consumption in developing countries has risen more than twice as rapidly as that in developed countries. However, population has also grown more rapidly in developing countries, reducing the difference in per capita growth. Indeed, per capita consumption in Africa declined over the last two decades, with population growing at 3.2 percent per annum, nearly 2 percentage points faster than milk output. Per capita consumption in Latin America rose in the 1960s, but stagnated in the 1970s and 1980s. Per capita consumption in Asia rose steadily at a substantial rate.

The "determinants" of milk consumption can be analyzed by country group or region to gain insight into plausible future trends. The major factors influencing the growth of milk consumption in any specific region are:

1. the growth rate of population,
2. the growth rate of real per capita income,
3. the income elasticity of milk demand,
4. any changes in the real prices of milk and other closely related goods,
5. any change in consumer preferences.

The expected effect of these factors is shown by:

$R = G_c - G_p - (G_y * E_{cy})$, where

R = Residual for region i, assumed to be caused by factors 4 and 5,

G_c = annual growth rate of milk consumption in region i,

G_p = annual growth rate of population in region i,

G_y = annual growth rate of per capita income in region i, and,

E_{cy} = the income elasticity of demand for milk in region i.

Given data on factors 1), 2), and 3), which are easily obtained, it is possible to imply the effects of factors 4) and 5). For example, a positive residual suggests that declining milk prices and/or taste changes have resulted in a more rapid rate of consumption growth than would have been caused by population or income growth (given the measured income elasticity), and vice-versa. The calculation for each major region is shown in Table 4.10.[10]

Table 4.10 Predicted vs. Actual Growth Rates of Milk Consumption, 1980-1988, by Region

Country Group/ Region	Income Growth Rate	Income Elasticity	Population Growth Rate	Predicted Consumpt. Growth Rate*	Actual Consumpt. Growth Rate	Residual*
Developed Economies	2.53	-0.05	0.56	1.15	1.87	0.72
Developing Economies	1.23	0.57	2.29	2.68	2.37	-0.31
Africa & Middle East	1.11	0.68	3.02	2.08	0.39	-1.69
Asia and Far East	2.28	0.52	2.01	3.45	3.70	0.25
Near East	1.77	0.53	2.16	3.12	3.14	0.02
Latin America	0.77	0.49	2.00	2.39	2.40	0.01

* Predicted consumption growth rate=income growth rate times income elasticity + population growth rate. Residual = actual less predicted consumption growth rate.

Source: Authors' calculations using data from FAO and World Bank.

The results show that population growth was a significant contributor to the growth of milk demand in every developing country region during the last three decades. In all developing country regions population growth was 2 percent per annum or higher and in Africa and the Middle East population growth was 3 percent. In contrast, in most developed countries population grew at less than 0.5 percent. In some West European countries population growth was negative.

However, per capita income growth was a relatively small contributor to the growth in milk demand in developing regions. Even in the area of most rapid per capita income growth, Asia and the Far East, rising income is estimated to have had only about half the effect of growing population. The calculated residuals are small, suggesting that income and population growth explain well the growth in milk consumption in most regions, with the exception of Africa and the Middle East where the growth in consumption falls well short of the predicted. Africa was unable to increase production at a rate meeting the potential increase in demand, and was also unable to obtain sufficient commercial or concessionary imports. There is some evidence that demand in Asia and the Far East has been expanding somewhat more rapidly than population and income would explain, and this is likely to be caused by a shift in tastes as milk is introduced to regions where it is not a traditional food.

When the data are further disaggregated, the growth of per capita income in developing countries shows wide variation. For the period 1961-1987, several countries in Asia registered high rates of growth, including 5.7 percent per annum for Hong Kong, 4.7 percent for South Korea and 3.3 percent for Malaysia. As a subregion, Sub-Saharan Africa registered a decline in the per capita income level and even in the level of real gross national product (GNP). Real per capita income did increase in North Africa and in the Middle East subregions. Income growth in Latin America also was low. With the exception of Argentina and Mexico, all the major countries in the region witnessed a growth rate of income of less than 1 percent per annum for the period 1961-87. Although real Argentine GNP grew at an overall rate of 2.8 percent for this period, the growth rate in the last decade has been only 0.06 percent. Of the 28 Latin American countries studied, 18 had negative real income growth for 1979-87. Low income growth was a major factor in the low growth of milk consumption (and production) in this region.

The growth rate of real GNP per capita in most European countries exceeded 2 percent during 1961-1987, and in the United States and Canada the corresponding figure was 1.9 percent. Nonetheless, overall demand for milk and butter has been stagnant or declining in all these developed countries. Japan is the exception, along with the former centrally-planned economies. The income elasticity of milk demand in most developed countries is very small and probably negative, implying that milk consumption will decline with a further rise in real per capita income. Nonetheless, the calculated residual in Table 4.10 is positive, suggesting either that prices have been declining or that consumers' tastes have changed toward a preference for more milk and milk products.

In the United States a similar decline in the consumption of dairy products was evident until recently. Demand declined 1.5 percent per year from the mid-1960s to the mid-1970s and then stagnated until the early 1980s when the trend was reversed. Per capita consumption of dairy products, excluding donations, increased at an average rate of 2 percent per year in the period 1983-86. A shift in consumer preferences towards dairy products seems to be responsible, with most of the increase occurring from greater consumption of cheese. According to the US Department of Agriculture, per capita cheese consumption grew more than 4 percent per annum from 1966 to 1986. The consumption of fluid milk and cream also increased slightly after a long period of decline.

It is unclear how far the recent upward trend in the U.S. consumer demand for dairy products will extend, or whether it will be reproduced in other developed economies. However, the recent increase in the demand for cheese and fluid milk and the decline in the demand for butter in the European Community seem to suggest that developments in the US market are not isolated phenomena.

As noted previously, average annual per capita consumption in the developing countries currently ranges from about 40 to 90 kg, while that in the developed countries ranges from about 250 to 700 kg. Although the range in individual countries is still greater, the difference between these two consumption ranges is broadly consistent with that which would be implied by the income elasticities used. For example, if per capita income grows at 4 percent annually, it will increase 120 percent in 20 years. Assuming an arc elasticity of 0.5 implies roughly a 60 percent increase in per capita milk consumption.

Two examples illustrate the results. First, an Asian country with a low per capita income, perhaps US$250, and a low level of milk consumption, perhaps 25 kg of milk, could be expected to increase these to about US$550 and 40 kg, respectively. Second, a Latin American country with an intermediate per capita income, perhaps US$2,000, and per capita milk consumption, perhaps 100 kg, would increase these to US$4,400 and 160 kg, respectively.

Each of these results seems broadly plausible. The first example extrapolates roughly to the second, and the second example extrapolates roughly to the situation of current developing countries. Thus, although only crudely indicative, these results are not glaringly incompatible with observed cross-section relationships between per capita income and per capita milk consumption. However, the assumptions are only plausible if it is also assumed that the income elasticity of milk demand drops rather abruptly toward zero once countries reach a per capita income level of about US$10,000.

A final note is in order. A huge variation exists in the level of per capita milk consumption, even among countries having similar income levels. See Table 4.11. This variation suggests that cultural factors have a great effect on per capita milk consumption. To the extent that such factors can change, they could be more important than income in determining future milk consumption patterns. There are recent signs, for example, that milk consumption in Japan and in China has been rising more rapidly than changes in income would imply given the expected income elasticities.

Table 4.11 Actual and Projected Milk Consumption in Selected Countries

Country	Per Cap Income (1987)$	Income Growth Rate	Per Cap Cons. 1960s (kgs)	Per Cap Cons. 1970s (kgs)	Per Cap Cons. 1980s (kgs)	Per Cap Cons. 2000 (kgs)	Per Cap Cons. 2010 (kgs)
Argentina	2360	0.0	196	205	194	194	195
Brazil	2020	2.8	77	86	95	111	128
India	300	1.4	40	42	53	58	62
Indonesia	450	2.3	1	1	1	2	2
S. Korea	2690	4.8	1	5	6	8	10
Mexico	1820	2.3	69	99	97	111	124
Sudan	330	0.0	92	92	134	134	134
Turkey	1200	2.8	134	121	105	124	142
Uruguay	2160	0.0	274	258	297	297	297

Notes:
(a) Per capita consumption are for periods 1961-70, 1971-79 and 1980-88.
(b) The income growth rates reported here are for the period 1961-88.
(c) The per capita consumption projections are based on the trend growth rate for the period 1961-1988.
Source: Authors' calculations using data from FAO Agricultural Production Tapes.

International Trade

The relevant figures for exports, imports, and net imports of condensed, dry and fresh milk by the study countries are presented in Table 4.12

Table 4.12 Exports and Imports of Condensed, Dry and Fresh Milk ('000 m. tons)

	Exports			Imports			Net Imports		
	'61-'69	'70-'78	'79-'87	'61-'69	'70-'78	'79-'87	'61-'69	'70-'78	'79-'87
Developing Countries	24	43	149	732	1200	2068	708	1157	1920
Asia	20	34	61	381	441	661	360	408	600
S. Asia*	–	–	–	27	40	90	27	40	90
E. & S.E. Asia**	–	–	–	247	270	410	247	270	410
India	–	–	–	41	37	39	41	37	39
Africa and Middle East	2	6	14	228	560	1002	225	554	989
N. Africa	–	–	–	89	191	349	89	191	349
Middle East	–	–	–	35	101	268	35	101	268
W. Africa	–	–	–	60	170	260	60	170	260
C. Africa	–	–	–	9	25	33	9	25	33
E. & S. Africa	–	–	–	41	73	92	41	73	92
Latin America	3	7	14	190	293	527	187	287	513
C. Am. & Carib.	–	–	–	80	139	280	80	139	280
S. America	–	–	–	109	155	247	109	155	247

(Continues)

	Exports			Imports			Net Imports		
	'61-'69	'70-'78	'79-'87	'61-'69	'70-'78	'79-'87	'61-'69	'70-'78	'79-'87
Developed Countries	1846	3664	6626	605	2184	4248	-1241	-1480	-2379
Japan	--	--	--	67	96	121	64	92	121
EEC	1173	2973	5541	529	2061	4118	-643	-912	-1422
W. Europe	61	88	133	35	51	65	-26	-37	-68
E.Europe	10	26	60	28	32	35	19	6	25
US & Canada	439	249	470	12	40	28	-427	-210	-442
Austr. & N.Z.	163	328	424	--	1	3	-163	-328	421

*excluding India
**excluding Japan

The blank spaces denote negligible amounts.

Source: Authors' calculations using data from FAO Agricultural Production Tapes.

Developed countries traditionally are the major exporters of milk products,[11] accounting for 97 percent to 98 percent of total world exports throughout the past three decades. Within this group, the European Community is the largest exporter, followed by the United States and Canada, and Oceania. Western Europe (excluding the European Community) exports marginal amounts, while Eastern Europe and Japan import significant quantities of milk but account for small shares of total milk imports.

In the 1970s and 1980s, the amounts exported by the European Community, Western Europe, and Oceania each increased by about 250 percent. The exports of the United States and Canada have remained constant over this period. The share of world trade enjoyed by the first three regions thus increased from about 67 percent to about 80 percent, with the share of the United States and Canada declining by similar amounts. Recent efforts by the European Community and the United States and Canada to control production have led to a sharp decline in accumulated stocks of milk products and it may be expected that their shares of exports will decline somewhat in the future.

Developing countries account for the bulk of milk imports, once intra-European Community trade is excluded. Regionally, the Africa and Middle East region has the greatest share of imports, followed by Asia and then Latin America. The major country importers in Africa and the Middle East are Algeria, Saudi Arabia, Libya, Egypt and Jordan. In addition to having the largest share of imports, the growth rate of imports is highest in the Africa and the Middle East region, followed by the Latin America region.

Asia's share in the total imports of developing countries declined from 52 percent in the 1960s to about 32 percent in the 1980s. The single largest Asian importer was Japan, averaging 121 thousand metric tons in the 1980s. The other major regional importers were the Philippines, Hong Kong and Singapore with 94,000, 66,000 and 60,000 metric tons, respectively.[12]

Latin America's share in total developing country imports has remained around 26 percent in the last three decades. The largest country importer in Latin America has been Mexico whose imports rose from an average of 30,000 to 173,000 metric tons over the three decades. Brazil is the next largest importer in this region, averaging 53,000 thousand metric tons in the 1980s.

Although substantial international trade occurs in milk products, the amounts traded are small in relation to production. Since trade flows are strongly influenced by government policy, prices in the international market are volatile and trading patterns are constantly changing.

Imports have risen where the increase in domestic demand exceeded the increase in domestic production. In some regions, as in the newly industrializing Asian countries, both production and demand have been rising rapidly. In other cases, as in most of Africa, milk production rose slowly and demand was encouraged by concessionary imports and food aid. Imports by North Africa grew at an annual rate of 14.6 percent between the mid 1960s and mid 1970s and at a rate of 9.2 percent between the mid 1970s and the mid 1980s. The corresponding rates of growth of imports by Middle Eastern countries were 20.6 percent and 18.5 percent.

Despite these rapid growth rates in imports, net imports remain small as a share of production for the developing countries, at least at the regional level. No region is heavily dependent on imports to maintain its consumption (Table 4.13). This pattern suggests that milk imports (and thus exports) are unlikely to become large relative to consumption. Milk consumption is thus likely to depend importantly on domestic production capacity, except for special cases.

Table 4.13 Net Imports of Condensed, Dry and Fresh Milk as Percent of Production in 1988, by Region

Region	Production (million m. tons)	Net Imports (million m. tons)	Net Imports as Percent of Production
Asia	68.6	0.7	1
Africa &			
Middle East	25.9	1.0	4
Latin America	39.3	0.5	1
LDCs	126.5	4.2	1
DCs	356.3	2.1	2

Source: Authors' calculations, using data from FAO Agricultural Production Tapes.

The World Dairy Industry: A Glance into the Future

Since projections assume the continuation of past trends, the results of the trend analysis are briefly summarized again. For production, growth rates have been most rapid in Asia, and accelerating, particularly in the sub-region of East and South East Asia. In Latin America, growth has been somewhat slower, and declining, with growth in Central America declining most. In Africa and the Middle East, offsetting trends are evident. In North Africa, output has risen and, although the growth in milk production has recently been somewhat slower, it remains high by international standards. In the Middle East, the growth rate has been moderate but declining. In the rest of Africa, growth rates are low.

In the developed countries, production grew at a much slower rate than in the developing countries, although there are some important regional differences. Because the production base in the developed countries was much larger, and because consumption was rising even more slowly than production, surpluses resulted which were large relative to total world milk trade. Efforts to control and dispose of milk surpluses induced surplus countries to impose production controls, reduce domestic prices, and subsidize exports.

Milk consumption showed a similar pattern to production, with the fastest pace of increase occurring in the newly industrialized countries in Asia and in the oil-rich countries in the Middle East and North Africa where per capita income grew rapidly. The growth rate of consumption was lowest in Sub-Saharan Africa, where per capita income declined for most countries during the 1970s and 1980s. In Latin America, a number of countries experienced a severe decline in per capita income during the 1980s, also leading to sluggish growth in consumption, although per capita milk consumption in this region remained the highest among developing country regions. Developed countries, as a group, experienced slow per capita growth of milk consumption, but several countries, including the United States, experienced a reversal of the historically declining trend of per capita consumption in recent years.

Milk production, consumption and international trade flows in the years 2000 and 2010 were projected under four different scenarios. The methodology, the assumptions underlying each scenario, and the corresponding results are discussed in turn.

Methodology

To examine the effect that rbST and MOET might have on future supply-demand milk balances internationally, efforts were made to identify past rates of growth of milk production and consumption and to project the values of these variables into the future using the rates identified. The trend rate of growth in milk production for each country was estimated by regressing annual production on an exponential trend, using data for the period 1961-1988. The resulting trend rate of growth then was used to project production to the years 2000 and 2010, using 1988 as a base production year.

A slightly different procedure was used to project milk consumption. The trend rates of growth of income and population for each country were estimated for the period 1961-1987 using the same procedure as for milk production.[13] These trend growth rates were then coupled with the estimates of *regional* income elasticities obtained from FAO in order to project consumption for each country to the years 2000 and 2010, using 1988 as a base consumption year. It was impossible to obtain estimates of specific income elasticities for each country in the sample. Because data on milk consumption are not reported, apparent consumption was calculated as production plus imports minus exports.

The difference between projected production and consumption constitutes the projected surplus or deficit in the years 2000 and 2010 for each country. The surplus or deficit for each subregion, region, and the world are obtained by appropriately summing the figures for individual countries. The surplus or deficit as a percentage of projected production is also calculated and used as a rough guide to potential market disequilibria. A more detailed discussion of the methodology adopted is contained in Appendix II.

Scenarios

Four scenarios are considered. Scenario 1 assumes that the trends observed from the period 1961-1988 continue to the year 2010. rbST and MOET are assumed to play no significant role in milk production. Scenario 2 assumes that the trends observed in the most recent period 1980-88 continue to the year 2010. rbST and MOET are assumed to play no significant role. Scenario 3 assumes that rbST and MOET are adopted in certain countries, leading to a one-time increase in the milk output of those countries. This effect is superimposed on the 1961-1988 trend rates of growth, as shown in Scenario 1. Scenario 4 also assumes that rbST and MOET are adopted in certain countries, with their effect imposed on the 1980-88 trend rates as in Scenario 3.

Scenario 1: If the trends evident over the period 1961-88 continue for milk production, population and per capita income, if the assumed income elasticities are valid, and if no other factors significantly affect consumption, it is projected that there will be an approximate balance between the supply and demand of milk in the year 2000, and a small surplus of about 3.6 million tons in the year 2010. (See Table 4.14).

Although the projections suggest approximate balance in world supply and demand, the same is not true at the regional level. In both 2000 and 2010, the developing countries as a group are net importers as they are today. The assumptions used in this scenario project a growing milk deficit for the developing countries, from roughly 6 percent of their total production in 2000 to 10 percent in 2010. Correspondingly, the developed countries have a growing milk surplus, from 4 percent of output in 2000 to 9 percent in 2010. The disequilibria at the regional and country levels would thus have to be resolved through substantial increases in international trade. In absolute magnitude, milk trade would have to expand roughly tenfold. Without such an increase in trade, there will be pressure for milk prices to

Table 4.14. *Scenario 1:* Milk Production, Consumption and Trade Balances Projected at the 1961-1988 Trend Rates

Trade Surplus (+) or Deficit (-)   in '000 m.tons

	2000		2010	
	S/D	%	S/D	%
Developing Countries	–10129	6	–22457	10
Asia	8108	8	13593	10
S. Asia*	-841	5	-3921	18
E. & S.E. Asia**	1460	34	4762	52
India	3514	6	3268	4
Africa & Middle East	-16700	51	-29489	73
N. Africa	-10031	93	-13571	95
Middle East	-3466	36	-8083	69
W. Africa	-779	47	-1504	81
C. Africa	-87	37	-164	55
E. & S. Africa	-2338	23	-6041	51
Latin America	2439	5	2924	4
C. Am. & Carib.	1197	8	2588	12
S. America	1243	3	336	1
Developed Countries	10351	4	26067	9
Japan	3974	33	9485	53
EEC	7979	6	21818	15
W. Europe	511	3	1277	7
E.Europe	1800	7	4302	14
US & Canada	–627	1	–2393	3
Austr. & N.Z.	689	5	1063	6
World	223	0	3610	1

*excluding India
**excluding Japan
Note: In the above table % refers to the percentage of projected production the surplus or deficit would be in the years 2000 and 2010. Components do not necessarily add to subtotals as data for some countries are not shown.

Source: Authors' calculations using data from FAO Agricultural Production Tapes.

rise in deficit countries and to fall in surplus countries.

The implied increase in the growth of milk traded internationally is substantially greater than that which occurred in recent decades, even including the increase encouraged by subsidized milk exports and food aid. For example, developing country milk imports rose at about 6 percent annually during the 1980s. If developing country milk imports were to continue rising at 6 percent to the year 2000, such imports would total about 4 million tons. In contrast, developing country imports of 10 million tons are projected under Scenario 1 (a 14 percent annual growth rate).

Although developing countries have a projected deficit in Scenario 1, that result masks large differences in behavior among developing countries. Of all the developing country regions, only the Africa and the Middle East region (including all of its subregions) is projected to have a deficit. Asia and Latin America are both projected to show surpluses, along with the developed countries (except for the United States and Canada which are discussed later).[14] Indeed, although the South Asia subregion has a small deficit, all other developing country subregions other than Africa and the Middle East have a projected surplus. Thus, the Africa and the Middle East region is projected to absorb nearly the entire surplus of the rest of the world. Eighty percent of the regional deficit is projected to be in the North Africa and the Middle East subregions.

The projections from Scenario 1 appear unrealistic insofar as milk imports in the year 2000 would account for roughly 90 percent of total milk consumption in the North Africa subregion. Few countries have been willing to accept such a high level of dependence on milk imports. A deficit of the magnitude projected seems likely to induce special efforts to increase production and/or decrease consumption, resulting in lower imports. Similarly, although projected imports would be a smaller percentage of consumption in the other Africa and Middle East subregions, the import shares are all high relative to those which have been considered politically acceptable by most countries in recent decades. The large deficits projected for East and South Africa, and also for West Africa, result mainly from rapid population growth. These deficits also seem large relative to the expected capacity of these subregions to import, suggesting that the projected imports might have to be concessionary.

Among developing countries, India has the largest projected surplus in the year 2000—3.5 million tons, followed by smaller surpluses in each of the other surplus subregions. The projected surplus in the Republic of Korea is 1.6 million tons, representing 47 percent of the country's projected production. Although income and population, and thus milk demand, are projected to grow rapidly, output is projected to out pace consumption, thereby converting Korea from a net importer to a net exporter.

The Indian and Korean cases raise problems which are repeated elsewhere in the projections. Even over a relatively short time period, the balance between projected surpluses and deficits can seem implausible. For example, the rapid increase in milk production in Korea has been encouraged by strong government financial incentives. It is unlikely that the same level of government support for the milk industry would be maintained if and when the country shifts from being an importer to an exporter. Accordingly, the projected surplus for Korea is unlikely to materialize. Similar questions can be asked regarding the projected Indian surplus, although India appears to have potential for developing a comparative advantage in milk under relatively free trade.

The Latin America region is projected to have a net surplus of 2.4 million tons, 4.5 percent of the projected regional output. In Latin America the three major producers, Brazil, Mexico and Argentina, are each projected to experience net surpluses, with Mexico having the largest surplus of 1.4 million tons. In contrast, Costa Rica, Chile, Peru and Columbia are projected to have small deficits of approximately 0.1 million tons each. Surpluses would thus have to be marketed outside the region or prices would have to decline sharply to achieve market equilibrium in surplus countries.

The developed countries, and the European Community in particular, were generally projected to continue with large surpluses. For Eastern Europe, acceptable data were only available for Hungary, Poland, and Yugoslavia.[15] These countries were projected to have a surplus of 2 million tons in 2000 and twice that in 2010. Even Japan was projected, using this methodology, to have a surplus of 4 million tons in 2000, 33 percent of the its total milk output in that year. And that projected surplus more than doubles in 2010. The United States and Canada was the only developed country region projected to have a deficit, although this was small, representing less than 1 percent of their output in the year 2000.

If the European Community maintains the production controls it has recently imposed, its projected deficit should be much lower than that projected. Removing the projected surpluses of India, the European Community, and Japan would go a long way toward maintaining balance, under the companion assumption that the projected imports by Africa and the Middle East are larger than seem plausible to finance through either commercial or concessionary means.

The projected surpluses and deficits in the year 2000 seem relatively balanced overall, but implausible in specific situations. The situation, in this scenario, is still less balanced in the year 2010. With the exception of India and the subregion of South America, the projected surplus or deficit of each region increases in the year 2010 relative to that in 2000. As projected, the surplus of India will decrease from 3.5 to 3.3 million tons, and the surplus of South America will decline from 1.2 to 0.3 million tons. In nearly every case, the trade imbalance also increases as a percentage of domestic production.

Scenario 2: Scenario 2 (Table 4.15) differs from Scenario 1 in that production and consumption are projected to grow at the rate estimated from the most recent sub-period 1980-88, rather than the longer period 1961-88. Use of the more recent period assumes that the recent past is a better predictor of future events than the longer trend.

Comparison of the data in Tables 4.14 and 4.15 reveals a number of important differences between the two projections. First, under Scenario 2, both developed and developing countries have a surplus of milk. The developing countries are projected to have a net surplus of 4 million tons in the year 2000, instead of a deficit of 10.1 million tons as in Scenario 1. This change results mainly from the higher rate of output growth during the 1980s. The surplus of the developed countries, on the other hand, falls from 10.4 to 7.2 million tons, reflecting policy efforts to reduce the growth of milk production in developed countries during the 1980s. The combined result of these changes leads to a world surplus of 11.2 million tons, 2.5 percent of projected production.

The greatest imbalance in production in Scenario 2 is in Asia, nearly all of which occurs in India. India is projected to have a surplus of 19.8 million tons in the year 2000, a large amount relative to its own production and relative to total world supplies. India alone has a projected surplus sufficient to supply all of the milk deficit regions. The Africa and the Middle East region has a large net deficit and is the only significant deficit region in the world. It seems unlikely that Africa will offer India an attractive export market.

As noted earlier, between 1979 and 1988, Chinese milk production and consumption rose at the rapid rate of 15.7 percent annually, but from a low initial base of about 1 million tons. The reported cumulative growth in production and in consumption were nearly identical over the period considered. A small deviation in either could produce a significant difference in future trade balances, but we expect China to remain mainly self-sufficient.

Latin America has a small net deficit instead of a net surplus. This reversal takes place mainly due to Mexico's large deficit; the last decade witnessed a considerable slowing of production while income and population growth remained largely unchanged. This situation could change relatively quickly if income growth resumes.

For the developed countries, Scenario 2 reverses the positions of the European Community versus the United States and Canada. The European Community, with the implementation of stricter supply control measures in recent years, has considerably slowed the growth rate of production. Continuation of this trend, along with the more recent increase in the

Table 4.15 *Scenario 2:* Milk Production, Consumption and Trade Balances
Projected at the 1980-1988 Trend Rates

(Trade Surplus(+) or Deficit (-) in '000 m.tons)

	2000		2010	
	S/D	%	S/D	%
Developing Countries	4005	2	18059	7
Asia	19830	17	45947	26
S. Asia*	-398	2	-3188	12
E. & S.E. Asia**	1571	37	5082	56
India	17700	22	41930	32
Africa & Middle East	-14604	45	-23740	59
N. Africa	-9541	86	-12308	82
Middle East	-2780	31	-6058	58
W. Africa	-790	49	-1504	84
C. Africa	-146	68	-321	131
E. & S. Africa	-1351	13	-3549	28
Latin America	-259	1	-2093	4
C. Am. & Ca.	-1964	19	-4473	40
S. America	1705	5	2379	5
Developed Countries	7222	3	19735	7
Japan	957	11	2054	19
EEC	-3366	3	-997	1
W. Europe	520	3	1295	7
E.Europe	-1200	5	1791	7
US & Canada	9283	11	17515	18
Austr. & N.Z.	1986	12	3714	19
World	11228	3	37794	7

*excluding India
**excluding Japan

Note: In the above table % refers to the percentage of projected production the
surplus or deficit would be in the years 2000 and 2010.

Source: Authors' calculations using data from FAO Agricultural Production Tapes.

growth rate of consumption, results in the projected deficit. In contrast, the United States and Canada are projected to have a considerable surplus. This is the result of continued income growth (recall that the income elasticity is negative) and a somewhat higher production growth rate in the last decade. In the developed countries the growth rate of population is low and thus has a relatively insignificant role. The surplus of Australia and New Zealand is also projected to increase in Scenario 2. Milk is an important commercial export for both countries. A deficit is projected for Eastern Europe, reflecting the downturn in production in the 1980s.

Under the conditions of Scenario 2, the supply-demand balance worsens considerably in the year 2010 relative to the year 2000 (Table 4.15). Both developed and developing countries, as groups, are projected to have larger surpluses, equal in each case to 7 percent of output. Each group is projected to have roughly a 50 percent share in world production by 2010. However, India and the United States and Canada are projected to have huge surpluses, while the Africa and Middle East region is projected to have a large deficit.

As noted previously, the surpluses and deficits resulting from projected production and consumption are very sensitive to minor changes in the growth rates of the relevant variables. The differences in the projections in Scenarios 1 and 2 highlight this point. Because milk supply is generally estimated to be more sensitive to changes in price than milk demand, especially in the short run, the projected world surplus is likely to result in a decline in the international price proportionally larger than the surplus (as a percent of world output). The surpluses outlined in Scenarios 1 and 2 are large relative to those experienced during the last decade, and accordingly could be expected to causes significant structural adjustment.

Part of the adjustment might occur via governmental efforts to protect producers from international price declines and to deal with surpluses via controls and stock accumulation, although the more likely response today would be milk price reductions. The projections strongly suggest that developing country regions like India, whose milk production has been increasing rapidly, are likely to witness slower growth of milk production in the future because demand will be a constraint.

Asia is the region with the most rapid economic growth during the last two decades. Thus, these projections already include the effect of past increases in income on consumption. However, if income in Asia or in other developing regions rises more rapidly than in the past, or if milk consumption rises more rapidly because of taste changes, the projected regional and world surpluses would be smaller. Because each of the three major developing country regions accounts for only a small proportion of world milk consumption, marginal changes in their consumption seem unlikely to strongly affect the projected world surplus. Even if more rapid consumption growth eliminated their own projected surpluses, the developed countries' surplus would remain.

Scenario 3: Scenario 3 assumes that rbST and MOET are adopted in certain countries, leading to a one time increase in the milk output of those countries which is superimposed on the trend rate of growth which is again, as in Scenario 1, calculated from the data for 1961-1988. For simplicity of design and exposition, and to avoid suggesting too much accuracy, the same adjustment is applied to output in 2000 and in 2010.

Percent Increase from Adoption of Two Biotechnologies

Ten percent: European Community, Western Europe, US and
       Canada

Five percent: Japan, Australia and New Zealand

One percent: Brazil, Mexico, Argentina, and India.

The assumed increases in milk production include the effects of both rbST and MOET, including any indirect cost savings caused by their interactions with other technologies. The total increase in milk in each case is assumed to come approximately 50 percent from rbST and 50 percent from MOET. The production increases due to the adoption of rbST and MOET are calculated as a percent of 1988 output. The increases are thus smaller than they would be if they were calculated as a percent of output in 2000 and 2010.

The assumed increases probably exaggerate the effects expected in 2000. For example, given the rather small expected unit cost savings from the use of rbST, as calculated using Perrin's approach (see Chapter 2), and the low price-elasticity of demand for milk in the developed countries, the adoption of rbST could not achieve an additional five percent increase in output in the developed countries unless a significant part of the increase could be exported. However, the projections suggest that developing countries are not likely to import significant milk, except for Africa which may not have the resources to pay. In addition, only a few years remain prior to the year 2000 for adoption to occur and a moratorium on the adoption of rbST is currently in effect in the European Community. In contrast, the assumed increases could be conservative relative to the potential in 2010, if the price of rbST declines significantly by that date, if the technology of its use has been further refined, and if the use of MOET begins to have a significant effect on the rate of genetic improvement.

The results of Scenario 3 are presented in Table 4.16. Not surprisingly, the projected world milk surplus is increased by the adoption of the two biotechnologies. Developing countries as a whole are projected to face slightly smaller deficits of 8.1 million tons in 2000 and 19.9 million tons in

Table 4.16   Scenario 3: Effect of Adopting rbST and MOET on Milk
Production, Consumption and Trade Balances, Projected at the 1961-1988
Trend Rates

(Trade Surplus (+) or Deficit (-)   in '000 m.tons)

	2000		2010	
	S/D	*%*	*S/D*	*%*
Developing Countries	-8127	4	-19853	9
Asia	9338	10	15316	12
S. Asia*	-839	5	-3921	18
E. & S.E. Asia**	1460	34	4761	52
India	4144	7	4087	5
Africa & Middle East	-16701	51	-29489	73
N. Africa	-10032	93	-13571	95
Middle East	-3466	36	-8083	69
W. Africa	-779	47	-1504	81
C. Africa	-87	37	-164	55
E. & S. Africa	-2338	23	-6041	51
Latin America	3810	7	4710	6
C. Am. & Ca.	1196	8	2588	12
S. America	2613	6	2121	4
Developed Countries	33204	12	50813	16
Japan	4574	36	10388	55
EEC	20832	15	36166	23
W. Europe	2219	12	3077	15
E.Europe	1800	7	4302	14
US & Canada	6900	8	5361	6
Austr. & N.Z.	1453	9	1909	11
World	25077	6	30961	6

*excluding India
**excluding Japan

Note: In the above table % refers to the percentage of projected production the
surplus or deficit would be in the years 2000 and 2010.

Source: Authors' calculations using data from FAO Agricultural Production Tapes.

2010. However, developed countries face much larger surpluses in each year, 33.2 million tons and 50.8 million tons under this scenario. Consequently, the total world surplus is projected to reach 25 million tons in 2000 and 31 million tons in 2010, 6 percent of projected world production in those years.

With higher output in India, Asia has a higher surplus. Latin America also has a somewhat higher surplus due to increased production in Brazil, Mexico and Argentina. Since it is assumed that rbST is not adopted (to significant degree) by any of the countries in Africa and the Middle East, the projected surpluses and deficits for that subregion remains identical to that in Scenario 1 (Table 4.14). With higher output in each of the developed countries, each subregion has a higher absolute projected surplus. In several cases, and especially in Japan, the projected surpluses are large relative to production suggesting substantial price pressure.

Scenario 4: Scenario 4 also assumes that rbST and MOET are adopted and with the same absolute effect as in Scenario 3, but this effect is now superimposed on the trends projected from 1981-88 which themselves suggest higher surpluses. The preponderance of surpluses across regions, the large surplus for each region except Africa and the Middle East, and the large world surplus in both 2000 and 2010, suggest that the milk prices would have to decline substantially. See Table 4.17. Conceivably, the world milk market would not exist as it is known today in the sense that trade would only occur under special arrangements between a number of surplus countries having no commercial outlets for their product and a number of deficit countries who are unable to pay. Whether this development occurs depends in large part on the strength of consumer preferences between fresh and reconstituted milk, since the former is traded internationally in only small quantities.

Summary of the Projections

Several prior studies have estimated or projected future milk output under varying assumptions regarding the effect of rbST adoption within individual nations. Although several of these studies suggested that the adjustment required as a result of rbST will be large, others argued that the expected milk production increases are not likely to be large relative to the change in production which would otherwise occur over time (e.g. Fallert et al., 1987). It has been argued that the output increases caused by rbST could be absorbed without undue adjustment cost, e.g., without a sharp decline in milk price and without provoking the exit of many of those farmers who today have higher production costs.

The projections in this study cannot determine whether the use of rbST

Table 4.17 Scenario 4: Effect of Adopting rbST and MOET on Milk Production, Consumption and Trade Balances, Projected at the 1980-1988 Trend Rates

(Trade Surplus(+) or Deficit (-) in '000 m.tons)

	2000		2010	
	S/D	%	S/D	%
Developing Countries	5993	3	29840	11
Asia	21101	18	47805	27
S. Asia*	-398	2	-3118	12
E. & S.E. Asia**	1571	37	5082	56
India	18517	22	43248	33
Africa & Middle East	-14609	45	-23740	59
W. Africa	-790	49	-1504	84
N. Africa	-9541	86	-12308	82
Middle East	-2780	31	6058	58
C. Africa	-146	68	-321	131
E. & S. Africa	-1351	13	-3549	28
Latin America	910	2	-763	2
C. Am. & Carib.	-1964	19	-4473	40
S. America	2875	8	3710	8
Developed Countries	31554	11	47195	15
Japan	1410	15	2594	23
EEC	8459	7	11267	8
W. Europe	2234	16	3105	16
E.Europe	-1200	5	-1791	7
US & Canada	17834	25	27324	25
Austr. & N.Z.	2817	16	4697	23
World	37546	8	77035	13

*excluding India
**excluding Japan

Note: In the above table % refers to the percentage of projected production the surplus or deficit would be in the years 2000 and 2010.

Source: Authors' calculations using data from FAO Agricultural Production Tapes.

or of MOET is likely to cause significant market disruption. At best, these projections are crude indicators. However, these projections cast some doubt on the sanguine view. In the developed countries, governments were hard pressed in the 1980s to avoid the accumulation of surpluses despite slower growth in milk output because consumption grew hardly at all. The projections undertaken in this study suggest that a tendency toward surpluses will continue in the developed countries even if rbST and MOET are not adopted, and these surpluses may also occur within many developing countries. If rbST and MOET add significantly to the growth of milk output, the situation will be exacerbated. Even a relatively small addition to the output of each of the major milk producers will increase future world surpluses in a compounding manner.

It might be argued that the projected production increases stemming from factors other than biotechnology will be lowered by the adoption of rbST and MOET, assuming that the latter causes a decline in price. That would not occur if the price of milk is supported, and it might not be valid even if the price of milk declines significantly. Production could accelerate because of a potential positive interaction between rbST, MOET, and other existing production-increasing technologies.

There is no argument here that adoption of rbST and MOET should be constrained. Economists generally advocate the adoption of cost saving technologies precisely because the gains to consumers are likely to outweigh any losses to producers. This case may be no different. However, it appears that in the foreseeable future the adoption of rbST and MOET will increase milk production much more in the developed countries than in the developing countries, requiring a significant price decline for market equilibrium. Developed countries might resolve the market adjustment problem via production controls at the individual producer level. This approach could leave individual farmers better off, as a result of lower costs and slowly rising production, but would bring reduced benefits to consumers. It is difficult to imagine that it will be politically feasible to make the implied adjustment in the developed countries without some accumulation of stocks, and these may again end up depressing world markets. However, there is some indication in the United States at least of a new-found political willingness to allow for substantial rationalization of the dairy sector, and the agreements reached under the Uruguay Round should encourage this approach.

This suggests that another perspective is worth considering. In 1988, each of the developing country regions had a net trade deficit in milk. The estimated trends indicate, however, that milk production has been growing more rapidly than consumption in all developing regions except Africa and the Middle East. The projections also suggest that many developing countries have potential to develop a significant comparative advantage in milk production in the next decade or two, despite the rapid increases expected in their populations and income per capita. The increased growth rate of milk production in developing countries reflects the diffusion and application of more traditional agricultural technology. This technology is appropriate for the resources currently available in the developing countries. Seen in this light, the development and adoption of rbST and MOET within developed countries may be a technology that, allows developed countries to remain competitive in milk production in the face of a rising challenge from developing countries whose production efficiency is rising. Correspondingly, milk costs may continue to decline worldwide, allowing milk to become increasingly accessible to the consumer even in poorer regions.

Notes

1. If, of course, rbST adoption causes a rapid increase in milk supply that encourages a shift toward greater market regulation, rbST will have much less effect on U.S. comparative advantage and conceivably could even slow the rate at which dairy sector productivity increases.

2. The United States' experience with rbST adoption in each of these aspects is likely to strongly influence subsequent legislation regarding rbST in the Economic Community, Canada, other developed countries, and probably developing countries as well.

3. The estimated growth rates for each country during each of three separate periods, 1961-70, 1971-79, and 1980-88, are available from the authors. They are not included here for reasons of space.

4. China is a relatively small producer and consumer of milk. In 1988, its per capita consumption was only 3 kg and its total annual production was about 3.7 million tons, less than 3 percent of less developed country milk production and less than 1 percent of total world production. Data were available for the growth of Chinese milk production and consumption only for 1979-1988 (F.C. Tuan, USDA). Due to data omissions, China is not included in the tables for Chapter 5, though the relevant information is included in the text. As China is a small producer, the information in the tables is significantly altered only where per capita milk consumption is discussed. Because of China's large population, its inclusion would reduce Asian per capita consumption by about 50 percent.

5. Milk production in China grew at about 15 percent annually from 1979 to 1988 from a low initial base.

6. Developed country milk yields rose at about 1.2 percent annually between the mid-1960s and the mid-1980s, while the average yield in developing countries rose hardly at all. Within both the developed and developing countries, the rate of increase varied greatly among countries and regions. For example, U.S. milk yields grew twice as rapidly as did Japanese milk yields, although each country provided high milk price support to its producers . Similarly, milk yields in South and South East Asia rose at 5 percent annually, while those in Latin America did not rise.

7. Latin America is the partial exception that proves the rule. Latin America contains large areas of extensive grasslands which are still only lightly exploited. The rising cost of land in areas of traditional milk production encouraged a shift into extensive grasslands to take advantage of low-cost pastures. This move was associated with a shift to the use of hardier dual-purpose animal breeds better able to cope in these challenging areas. The technological/geographical shift resulted in a lower cost of milk, but also lower average milk yields, thereby masking the effects of technological and management progress that have been occurring within the region.

8. Because consumption data are not reported by FAO, "apparent" consumption is calculated as production plus imports, less exports. Changes in inventories, which are important in some years for some countries, have not been incorporated into these calculations.

9. The income elasticities used in Table 4.10 range between -0.05 for the developed countries as a group to about 0.6 (0.5 to 0.7) for the developing countries. The income elasticity of milk demand appears fairly constant over a substantial range of lower income levels but appears to decline sharply at high levels of

income. It is not clear at what income level the decline begins. The elasticities are utilized in Table 8 as if they were arc elasticities, i.e., constant over the entire range of income changes assumed, an assumption that could exaggerate the expected increase in milk consumption as income increases. However, most developing countries are not likely to reach a high income level within the period of projection, so the bias should not be large.

10. Trade in cheese and butter occurs almost wholly among developed countries and is not discussed in this chapter.

11. China is essentially self-sufficient in milk production, importing and exporting less than 1 percent of total consumption.

12. The data set included one less year for these series.

13. Regression analysis using the data available for consumption and production suggested that the growth of production exceeded consumption, resulting in a projected milk surplus for China in 2000 and 2010. This result seems to be a statistical artifact. Given the structure of China's economy and Chinese economic policy, China is unlikely ever to export significant milk products. Neither is it likely to import significant amounts of milk in the foreseeable future.

14. We considered introducing projections for other Eastern European countries and for the ex-Soviet Union, but decided against it. The available data are not reliable. Recent political and economic changes suggest that projections from past data would have little value in any case. It is unclear how milk output, consumption and trade will unfold.

5

Summary and Conclusions

Technological change, brought about by the application of science to industry, is a prime engine of modern economic growth. The development of biotechnology is but one of the latest major types of technological innovation that has the potential to achieve a significant increase in economic well-being. However, while new technologies have the potential to reduce costs and improve quality, they may prove more profitable in some contexts than others, thus causing a shift in productive structures across regions and countries, in factor incomes, and in product prices. The net effects of technologies are complex and difficult to predict.

This study analyzes the likely effects of two new biotechnologies, rbST and MOET, that are applicable to the dairy industry. The analysis examines the potential for their profitable adoption in different contexts and focuses on the extent to which their effects may be different in developed and developing countries. The analysis suggests that these two biotechnologies are likely to be adopted eventually in developed and developing countries, but that each technology will be adopted earlier in developed countries. In the case of those developing countries which currently have the least favorable production environments, the adoption lag may be decades.

Adoption of the new technologies will result in lower costs of milk production and should result in lower milk prices. Prices should fall faster and probably further in developed countries during the next decade or so. To the extent that the international milk market is integrated across countries, consumers throughout the world will gain. Producers that are able to profitably adopt the two technologies will generally also gain. Those producers who cannot adopt are likely to lose as the two technologies are adopted by others.

rbST

The adoption of rbST has begun in the United States, where it can be expected to increase milk output, reduce milk prices, and increase the rate of structural change in the dairy industry. The use of rbST does not appear to produce substances that are harmful to humans or to the animals in which it is injected, though any conclusion is subject to further scientific study that will occur after rbST has been in use for some time.[1]

Dairy farms with relatively better management, higher yields, and/or larger herds should profit most from rbST and therefore are likely to be the first adopters. Using Marion and Will's (1990) model, we estimate that adopting farms with these characteristics have the potential to significantly increase farm profit. It is difficult to imagine that rbST will not be adopted rapidly by the owners of such herds. If so, given the larger than average size of such farms, national milk output may increase even more than proportionately to the number of farms adopting.[2] However, the results of recent surveys in California show that adopters are generally using rbST on only a small fraction of their herd, apparently because many animals do not respond sufficiently strongly to justify the cost of rbST.

The price of milk could decline significantly as the adoption of rbST accelerates.[3] If no human health problem is traced to rbST, its use should spread fairly rapidly as competition will spur adoption. As Perrin (1991) has noted, however, the equilibrium change in the milk price will be significantly less than the increase in output per animal. Using Perrin's analytical framework, but with calculations using parameters from the relatively more efficient dairy firms who will increasingly dominate the dairy industry, the unit cost of milk is estimated to decline by about 7 percent.[4] Given that the price elasticity of demand in the United States is low, milk consumption will increase by only about 2 percent, which suggests that unless sizable quantities of milk can be exported, consumption too will increase little in response to the price decline.

The estimated income elasticity of milk in the United States is also low (zero or even negative). From 1970 to 1990, milk consumption per capita fell slightly; fluid milk and cream consumption declined by more than 10 percent, butter consumption remained constant, and cheese consumption nearly doubled. During the same period, the average amount of milk produced per cow rose from about 10,000 lbs to more than 14,000 pounds, while the number of dairy cows decreased from about 12 million to about 10 million. During this period, U.S. public policy focused on sustaining the incomes of dairy farm families as the shift in milk supply outpaced that of demand. If rbST is adopted, increasing production and reducing milk costs, the number of dairy cows will decrease further as well the number of dairy farms. Taken on that criteria alone, a rapid increase in supply due to the use of rbST may not appear socially attractive insofar as it will cause many dairy farms to exit the industry more quickly than would otherwise have occurred.

Nonetheless, the total benefits of adoption should amply exceed total costs. Benefits will occur to producers who can profitably adopt, particularly early in the process,[5] but most benefits will be captured by domestic consumers. Even this relatively small decline results in substantial consumer benefits. Assuming milk consumption of approximately 70 million metric tons in the United States, an eventual 7 percent reduction in the farm level cost of production (a decline of approximately $0.02 per kg), and constant processing and distribution margins, it is estimated that consumer benefits will increase by about $1.4 billion annually.[6]

The use of rbST has been approved in the United States, but not in most other developed countries. Introduction of rbST in Canada and in the Economic Community has been stopped by concern that adoption of rbST might produce such an increase in milk production as to sharply reduce market price, requiring painful adjustment among dairy farmers, and/ or that its use, directly or indirectly, poses a threat to human health. However, if the U.S. market absorbs rbST without a politically unacceptable disruption of the milk market and if use of rbST results in no major human or animal health problems during its first several years of adoption, it seems likely that most other developed nations will also approve adoption. The delay in adoption among other developed countries offers potential for the U.S. to increase its comparative advantage in milk production, at least temporarily. Some countries, such as Germany, may continue to resist the introduction of rbST given those countries' deeper aversion to the concept of biotechnology. However, it seems likely that if their competitive position in the world milk market erodes, there will be pressure for approval of the use of rbST.

The use of rbST will not be profitable in most parts of most developing countries in the short to intermediate run. Milk production in most developing countries is carried out within a much less favorable environment than exists in developed countries. Climate is often more harsh. In tropical areas, temperatures and humidity are higher. The disease-parasite challenge is greater, subjecting animals to stress and reducing their ability to produce high milk yields. Producers have fewer management skills, particularly as regards the use of technologies like rbST that require improved animal husbandry and precise adjustment of other inputs, especially nutrition, for profitability. Farmers in developing countries also often face lower producer prices, caused in part by a lack of transportation infrastructure and inefficient milk collection, processing and distribution systems. The use of rbST will increase production risks along nearly every dimension of the production system. This higher risk will also discourage producers from adopting rbST, particularly in developing countries where credit markets function less well.

The cattle currently used in many developing countries, primarily *Bos indicus* breeds and their crosses with *Bos taurus* breeds, have been selected for their hardiness. The response of these cows to rbST will largely determine whether rbST can be used in developing countries in the intermediate future. Evidence from trials in Zimbabwe of rbST on Bos indicus and crossbred cows (Phipps, et al., 1991) suggests that such cows may respond better to rbST than was initially believed, provided that management is good and the production environment is favorable. Nonetheless, the base yields of the lower yielding cows are so low that even a high percentage response results in too small an absolute increment to make use of rbST profitable under the most optimistic assumptions, given the existing price of rbST.[7]

If the price of rbST declines when current patents expire, the use of rbST may be possible even on cows with relatively lower yields in developing countries.[8] In the short to intermediate run, however, the great bulk of developing country dairy producers will not find it profitable to make the complex set of complementary adjustments to their production system (nutrition, management, veterinary medicine, and genetic development) that are required for profitable use of rbST even were its price to fall significantly.

Looking at the complementarities between rbST, as a new technology, and existing factors of production, it seems clear that the use of rbST will increase the return to management in favorable production conditions. Thus, rbST is a technology that is more complementary in the short to intermediate run with management in developed countries than in developing countries. It is still unclear how the use of rbST will affect other factors, like land, since it appears that the higher feed efficiency achieved by rbST could reduce overall feed demand. However, the demand for certain types of quality feed may increase.

To the extent that the introduction of rbST reduces the unit cost of milk production in developed countries more than in developing countries, this new biotechnology will improve the competitive position of the developed countries. Potentially, developing countries could suffer reduced welfare. However, the process of technological change is not a zero-sum game. Technological change in milk production offers potential gains to developing countries in the same sense that cheaper Japanese automobiles and electronic goods offer benefits to U.S. consumers. These gains could be particularly large in areas like Africa, where there is prospect that the growth in future milk demand will greatly outstrip the growth in regional milk supply. Lower cost imports from developed countries have potentially important income and nutritional effects.

In the longer run, the use of rbST may be feasible even in developing countries and perhaps on a range of livestock types that is unexpectedly wide. This depends on the extent to which competitive conditions induce a decline in the price of rbST, and whether ongoing, broad scale rural development leads to improved management skills, nutrition, herd genetics, herd health and a marketing infrastructure. Improvements of this sort are precisely the developments that are needed to achieve significant gains in milk production even in the absence of rbST.

The rapid rates of increase in milk output that have been achieved in developing countries during the last decade have resulted from the adoption of "traditional" technologies that have been available for some time. Continued adoption of such technologies is highly suitable to these countries' current resource endowments and exploitation of the stock of "traditional" technologies offers great potential for these countries to achieve higher milk output at declining cost. The existence of rbST should not distract developing countries from this path. Since incorporation of these technologies is fundamental for profitable use of rbST in the future, it should enhance their importance and the emphasis on bringing them about. Further, they may be able, by use of such technologies, to reduce milk costs much faster than developed countries can do via the adoption of technologies like rbST.

MOET

The use of Multiple Ovulation and Embryo Transfer (MOET) is now practiced commercially in the United States and its use is increasing throughout developed countries, though mainly on an experimental basis. MOET has been used in developed countries primarily to increase the intensity of genetic selection. MOET makes a greater contribution when utilized in breeding programs in conjunction with artificial insemination because the possible intensity of selection among males is still much greater than the intensity of selection among females.

Under current practice, use of MOET is increasing the rate of genetic improvement in developed country dairy herds, though its effect is still relatively small, probably less than 1 percent per decade in the aggregate. This rate will increase as MOET becomes more economical and its use more widespread, though the use of MOET is not likely to become a general practice because it is only marginally cost effective.

The use of MOET will be even less profitable for use in breeding programs in developing countries than in developed countries, particularly in the short to intermediate run. The use of MOET as part of an on-farm breeding system requires a high level of on-farm management expertise and labor skill, as well as an efficient communication and transport system to permit timely interaction with technical specialists. These requirements are commonly lacking in developing countries. Indeed, the implementation of artificial insemination has produced poor results in many developing countries, in part because it has proven difficult to implement adequate recording systems without which the desired genetic selection is impossible. The problems facing the use of MOET will be similar but even greater.

It may be economical to use MOET within an Open Nucleus Breeding System (ONBS), wherein artificial insemination and MOET are combined to achieve a higher rate of genetic improvement within a small nucleus herd. There is potential for significant upgrading of the milk potential of many breeds found in developing countries. Improvement begins with an already genetically-superior nucleus herd of local cattle, and the improved progeny are released for use in the national herd. Little evidence is yet available regarding the actual intensity of selection that can be achieved within a developing country context, though it is expected to be significantly lower than the theoretical potential. It thus seems unlikely that MOET will make a large contribution during the next decade to the intensity of genetic selection in developing countries. It is likely to have greater effect in the longer run, though it is unclear whether it will prove more economical to upgrade the local herd through breeding programs using MOET and artificial insemination or to introduce exotic genes through cross-breeding.

MOET has potential for use in crossbreeding to produce a continuous flow of F1 generation cows under some conditions.[9] The available evidence suggests that F1 generation cows enjoy a significant milk yield differential relative to that of F2 generation cows and may also have better reproductive performance. This performance advantage make F1 generation female calves more valuable than F2 generation female calves. Although natural service or artificial insemination cannot be used to produce F1 calves out of F1 cows, MOET can achieve F1 generation calves. The question is whether embryo transfer can be implemented sufficiently cheaply in a developing country context to make the continuous production of F1 generation calves economical.

In fact, the use of MOET, appears too expensive at a cost of $250 to $300. However, American Breeders Service has developed a process to produce cheap in vitro generic F1 embryos using eggs obtained from ovaries recovered at slaughterhouses in the United States which are subsequently fertilized using semen from bulls in developed countries. These generic F1 embryos can then be transferred to recipient cows in developing countries at a total cost that may approximate $100. The calculations undertaken in this study suggest that the differential value of an F1 female calf versus an F2 calf, male or female, appears to be sufficiently great in some developing country contexts to justify the use of embryo transfer for this purpose. However, the profitability of this approach is sensitive to the assumptions made about the F1-F2 milk yield and reproductive differentials and the success rate of embryo transfer. Considerable experimentation will be required to determine whether the process is economical.

rbST, MOET and World Dairy Trade

A review of recent trends in world milk production, consumption and international trade indicated that the growth rate of milk production has been declining in both developed and developing regions (unweighted country averages) during the last three decades. Although it is well known that milk production has been growing more slowly in developed countries, where income elasticities are very low, it had been expected that the rate of milk consumption was growing more rapidly in developing countries due to rapidly rising population and per capita incomes. However, income growth was insufficient to fuel much growth of milk consumption in developing country regions during the 1980s, except in Asia.

Although growth rates of milk production have been declining in most of the world (except Asia), those in developing countries have systematically exceeded those in developed countries during the last three decades.[10] In developing countries, expansion of the number of dairy cows and dairy farms, the adoption of conventional dairy technologies, and gradually improving management have all contributed to rising milk production. Milk demand in developing countries is expected to grow in the future at a higher rate given a high rate of population growth and a higher growth of per capita income. Even without adopting rbST and MOET, developing countries should be able to increase the rate of growth of milk production because there is a sizable stock of conventional technologies that are profitable and that have not yet been adopted by dairy farmers.

Most developing countries in Asia and in Latin America also have created institutions that should be able to sustain a high rate of technological improvement in the milk sector. Thus, even if the growth of milk demand increases in developing countries, rising demand should be met by rising domestic supply, with relatively little scope for increased imports from developed countries. The potential for increased supply is lower in Africa and the Middle East, where resource and institutional constraints on domestic production remain severe. Africa also lacks income to purchase larger milk imports and will continue to rely on concessionary imports for an important share of its milk.

In developed countries, management improvements and technological change continue to increase dairy efficiency. These improvements have sustained a downward pressure on milk prices and encouraged consolidation of the dairy sector. Milk price supports, production quotas, and export subsidies were used throughout the 1980s to ease price pressure on domestic farmers.

As a result of export subsidies from developed countries, international milk prices fell dramatically during the 1980s. Because many developing countries are net importers of milk products, developed country milk export subsidies may have improved the welfare of developing countries. However, in most developing countries policies have been designed to improve producer rather than consumer welfare. Most developing countries increased the level of protection for domestic dairy farmers as international milk prices fell.

As the budgetary costs of domestic milk policies rose, milk producing developed countries attempted to rationalize policy. Significant progress was made in this regard, including an international agreement to restrict the use of price supports and export subsidies in the recent Uruguay Round of trade negotiations. These policy changes have reduced the domestic milk prices and milk stocks in developed countries and have slowed the rate of growth of milk output. As future technologies reduce costs, these effects are now more likely to be reflected in lower prices than in rising stocks and greater export subsidies.

Only a small proportion of world milk production is traded internationally and small changes in production or consumption of major milk producers can significantly affect the projected trade balance. The European Community, the United States and Canada, and Oceania are each significant exporting regions. Eastern Europe and Japan import significant quantities, but account for only a small share of total milk imports. Developing countries account for most milk imports. Among developing country regions, Africa and the Middle East has the largest share of milk imports, though much is imported on a concessional basis. Asia and Latin America each also account for significant milk imports.

Net milk imports remain small as a share of production in all developing regions, even in Africa and the Middle East, and it is likely that developing countries will remain relatively self sufficient in milk. Producer lobbies are strong, consumers have a preference for fresh as opposed to reconstituted milk, and the cost of transporting fluid milk is high relative to its value. Thus, significant segmentation of milk markets will continue, although trade in milk products will be sufficient to reduce the price differentials that would otherwise exist.[11]

The growth rates of production, consumption, and international trade in milk products was projected into the future under several different sets of assumptions. These projections suggest that world supply is likely to increase systematically more rapidly than world demand during the next decade or two.

If the trend rates of growth of the last three decades are used to extrapolate into the future, the world is projected to have an approximate balance between the supply and demand of milk in the year 2000 and a small surplus of about 3.6 million tons in the year 2010. In both cases, developed country regions remain net exporters, while developing country regions remain net importers. However, this result masks important regional variations. Among all regions, only the Africa and Middle East region has a projected deficit. That region is projected, under the assumptions utilized, to absorb the entire surplus of the rest of the world. Africa and the Middle East cannot realistically import so much milk except on concessionary terms.

The supply-demand situation is still less balanced if the trend rates of growth from the past decade are used for extrapolation. Production growth rates have been somewhat higher relative to consumption growth rates in the recent past in the most important milk producing countries. Under this scenario, therefore, much larger milk surpluses are projected, especially in India and the United States and Canada. These projections suggest that the market is likely to equilibrate via the imposition of production controls or via significant milk price declines in the countries with surpluses.

Additional projections were carried out to examine the effect of adopting rbST and MOET in a number of countries. These additional projections assume that the same trend rates of production can be maintained using conventional technologies other than rbST and MOET, so that adoption of these two biotechnologies would achieve an additional increase in milk supply. The expected increase in milk output from the adoption of rbST and MOET is thus added to the amounts projected using past trend rates. Not surprisingly, since the surplus in the United States and Canada is projected to increase even without adoption of rbST and MOET, adoption of these new biotechnologies further increases the projected surpluses.

Projections of future milk trade surpluses and deficits are highly sensitive to relatively small changes in the assumed trend growth rates of production and consumption. Milk consumption is sensitive to changes in income in developing regions, though not in most developed regions. The 1980s was a period of slow growth, particularly in Africa and Latin America. Per capita income growth could accelerate in the developing regions in the future and, if so, would reduce the projected trade surpluses. An acceleration of world growth by one full percentage point relative to the growth rates assumed in the projections would reduce most of the surplus projected for developing countries in Scenario 4, though a large surplus would still be projected for developed countries.

Several previous studies of the potential effects of the adoption of rbST on future milk output and dairy sector structure in the United States have concluded that adoption of rbST could cause a large increase in supply, requiring large industry adjustments, while others have suggested that the effect of rbST will be small relative to the effects of other conventional technologies that will also be adopted in the future. The results of this study suggest that the industry adjustment required in the United States following adoption of rbST, and MOET, are likely to be large.[12] If the adoption of other technologies shifts the supply curve outwards, these two technologies will shift it still more. More importantly, the additional shifts in supply will doubtlessly be more disruptive in the sense of requiring more sectoral adjustment at a more rapid pace.

The availability of rbST and MOET will increase the return to management skills and will probably reward scale. As a result, the dairy industry will contain fewer but higher yielding cows, fewer but larger dairy herds, and fewer but more skilled dairy farmers. Ironically, however, the spread of conventional technology in developing countries will bring about the same phenomena — increased specialization, commercialization and consolidation. Improvements in transportation and energy infrastructure, and in processing and distribution systems in developing countries will reinforce the process. Eventually the developing countries can be expected to adopt rbST and MOET as part of a continuing effort to reduce costs. The timing of their adoption in developing countries will depend primarily on the degree to which improvements are made in the production system (management, animal nutrition, herd genetics, and the disease-parasite challenge) in the interim, as well as the cost of rbST and MOET. Most developing countries are achieving improvements in the production system and potentially great improvements can be achieved over the longer run.

An increase in the rate of technological change in the dairy industry offers considerable social benefit. In the case of rbST and MOET, most of the benefits should accrue to consumers through declining prices, though some will accrue to producers, especially those who adopt early (Lemieux and Wohlgenant, 1989). Given the expected distribution of benefits from the adoption of these two new technologies, there is ample justification for programs that speed and ease the costs of the dairy sector adjustment that their adoption will also encourage.

In developing countries, any resulting decline in the price of milk may have special nutritional benefits. Recent research suggests that child nutrition in many developing countries is significantly and positively associated with the amount of animal products in the diet. Milk can be consumed easily by children, though its relatively high cost makes it economically inaccessible to many of the world's poor. As incomes rise and the price of milk falls, milk consumption will increase, and higher consumption can be expected to contribute positively to improving child nutrition.

Notes

1. For example, there is still concern that use of rbST could cause increased levels of mastitis in cows, and that the need to treat mastitis could increase the level of antibodies in milk – threatening human health. However, tight controls are maintained on the level of antibodies in milk, so this should not occur.

2. Monsanto claims that the number of adopters had reached 13,000 in December 1994 (New York Times, 1994), with a 50 percent increase in the prior six months. This is about 11 percent of the nations 120,000 commercial dairy farmers.

3. Data released just as this manuscript was completed indicates that during the first year rbST was available for adoption in the United States, the price of milk declined 9 percent, from $12.50 in December 1993, to $11.38 in December 1994. It is unclear whether this decline was due to an increase in milk production caused by the rising use of rbST or to unassociated factors. Milk consumption rose by 0.8 percent.

4. This calculation assumes that rbST is priced at $0.30 per daily cow dose. If the price or rbST falls to $0.05 per daily cow dose after expiration of patents, the price of milk could fall by more, nearly 9 percent.

5. The manufacturer of rbST, seeking an attractive return on its investment in research and development, has established a relatively high price for a unit dose of rbST. That price, which is significantly higher than the marginal cost of production, sharply reduces the on-farm profitability of rbST. In the short run, this means that adopters will gain a somewhat smaller benefit relative to nonadopters than if the price of rbST were lower.

6. In 1994, world milk production was approximately 550 million tons. Assuming an eventual saving of $0.02 per kg. and no increase in the consumption of milk, total annual savings to consumers worldwide from use of rbST would be about $11 billion. If the use of rbST is constrained to the major developed country milk producing regions, i.e., the European Community and Western Europe, the United States and Canada, and Oceania, which produce about 45 percent of milk worldwide, the total annual savings would be about $5 billion. Although terribly crude, these estimates provide some magnitude to the potential benefits from the new technology.

7. The trials suggest that the percentage increase in milk output achieved is negatively related to the base yield of these animals, i.e., cows with lower yields increase their daily output in response to rbST in greater proportion than do cows with higher yields.

8. Although the arithmetic calculations indicate a potential for profitable application of rbST conditions might support use of rbST, management will upgrade herd genetics prior to or simultaneously with adoption of rbST. Thus, it is not likely that rbST will be used on low yielding cows.

9. The offspring of two purebred animals is a first generation (F1) crossbred having an equal number of genes from each parent. If two F1 crossbred animals are bred, their offspring is a second generation crossbred (F2) having the same mix of genes.

10. Developed countries currently account for nearly 75 percent of world milk output. The share of developing countries has been gradually rising.

11. Although the two new biotechnologies analyzed in this study will improve the comparative advantage of developed countries, the effect on international markets is likely to be relatively small. Developed countries seem likely not to rely on the large milk export subsidies that appeared during the 1980s. With the moderate levels of protection that are still permitted developing countries, and a preference for fluid milk, most developing countries are unlikely to greatly increase milk imports.

12. The adoption of MOET could have a significant long-term effect on the unit cost of milk and thus on the economic structure of the dairy industry, yet the introduction of MOET has attracted little attention and no broad resistance, in contrast to rbST.

Appendix 1:
List of Study Regions and Countries

Developing Regions and Countries

Asia
  South Asia
    Bangladesh
    Pakistan
    Sri Lanka
  East and South East Asia
    Fiji
    Hong Kong
    Indonesia
    Republic of Korea
    Malaysia
    Papua New Guinea
    Philippines
    Thailand
    Samoa
  India

Africa and the Middle East
  North Africa
    Algeria
    Egypt
    Libya
    Morocco
    Sudan
    Tunisia
  Middle East
    Israel
    Jordan
    Kuwait
    Saudi Arabia
    Syria
    Turkey
    United Arab Republic
    Yemen Arab Republic

Yemen Democratic Republic
West Africa
Burkina Faso
Benin
Chad
Côte D' Ivoire
Gambia
Ghana
Guinea-Bissa
Liberia
Mali
Mauritania
Niger
Nigeria
Senegal
Sierra Leone
Togo
Central Africa
Burundi
Cameroon
Central African Republic
Congo
Gabon
Rwanda
Zaire
East and South Africa
Botswana
Ethiopia
Kenya
Lesotho
Madagascar
Malawi
Mauritius
Somalia
South Africa
Swaziland
Tanzania
Uganda
Zambia
Zimbabwe
Latin America
Central America and the Caribbean
Antigua

Bahamas
Costa Rica
Dominican Republic
El Salvador
Grenada
Guatemala
Haiti
Honduras
Jamaica
Mexico
Nicaragua
Panama
St. Vincent
Trinidad and Tobago
South America
Argentina
Bolivia
Brazil
Chile
Colombia
Ecuador
Guyana
Paraguay
Peru
Surinam
Uruguay
Venezuela

Developed Regions and Countries

Australia and New Zealand

Europe
Eastern Europe
Hungary
Poland
Yugoslavia
European Community
Denmark
France
Germany, Federal Republic
Greece
Ireland

Italy
Netherlands
Portugal
Spain
United Kingdom
Western Europe
Austria
Finland
Iceland
Malta
Norway
Seychelles
Sweden
Switzerland
Japan

United States and Canada

Appendix 2:
Data and Methodology for Projections

Data

The main sources of data utilized are:

1. population and GDP for individual countries, 1961-1987, *The World Tables 1989*, The World Bank, 1989,

2. production of milk and number of milk animals for individual countries, 1961-1988, *FAO Production Yearbook*, standard tape 1989, FAO,

3. export and import of milk for individual countries, 1961-1987, *FAO Trade Yearbook*, standard tape 1989, FAO, 1989,

4. estimates of income elasticities for milk for individual countries, FAO, and

5. other data are cited explicitly in the text when discussed.

Methodology

Output Projections

In Scenario 1, to obtain an estimate of the trend growth rate of milk output for each country, a semi-logarithmic trend equation was fit to the annual production data for the period 1961-88. This approach followed the simple model described in equations 1) and 2):

(1) $Y_t = Y_0 (1+r)^t$, where

Y_t = estimate of the output in the year t,

Y_0 = estimate of the output in the base year,

r = annual trend growth rate of output.

Taking logs of both sides of (1) yields:

(2) $\ln Y_t = \ln Y_0 + t \ln(1+r)$, or

$\ln Y_t = a + bt$, where

a = log of the base year level of output,

$b = \ln(1+r)$.

Equation (2) was fit to the production data for the period 1961-88 for each country. The estimated coefficients, $\hat{b}$, were used to calculate the estimated growth rates of production for each country by using the relation:

(3) $\hat{r} = e^{\hat{b}} - 1.$

The estimated growth rate, $\hat{r}$ was used to project output in the terminal years, 2000 and 2010, using the following relation:

(4) $Y_T = Y_{1988}(1 + \hat{r})^{T-1988}$, where

   $T = 2000$ or 2010,

   Y_T = projected output in the year 2000 or 2010, and

   Y_{1988} = output for the base year of 1988.

The projected milk production in the years 2000 and 2010 for each of the countries were then aggregated to give the projected output for each of the regions and subregions.

In Scenario 2, the method of projection is identical except that in estimating $\hat{b}$, data were used from the period 1980-1988 instead of 1961-1988. In Scenarios 3 and 4, the projected outputs in the terminal years were inflated by 1 percent, 5 percent, or 10 percent of 1988 output for the countries assumed to adopt rbST and MOET, with the percent depending on the authors' judgment of the likely degree of adoption.

In using equation (4) for projecting output for the individual countries certain exceptions were made. First, for countries which showed a negative estimated growth rate (i.e., $\hat{r} < O$), it was assumed that no further decline in output will take place after 1988. The projections thus assumed that future output equaled output in 1988. This assumption had important effect in Scenario 2 because a number of countries in the Sub-Saharan Africa and Latin America subregions experienced a decline in output for the period 1980-88.

Secondly, some countries in Asia, like the Republic of Korea and Indonesia, had extremely rapid growth rates, i.e., 20 to 25 percent per annum. These growth rates were believed to be the result of the very low base levels of recorded output when the period began and were considered unsustainable in the future. In these cases, a ceiling of 8 percent on the annual growth rate was imposed in the projections.

Consumption projections

Following the procedures outlined above it was assumed that the projected per capita consumption in the terminal year T was given by:

(5) $C_T = C_{1988} (1+i*m)^{T-1988}$, where

$\quad$ T $\quad$ = 2000 and 2010,

$\quad$ C_T $\quad$ = estimated per capita consumption in the year T,

$\quad$ C_{1988} = estimated per capita consumption in the base
$\qquad\qquad$ year, 1988,

$\quad$ i $\qquad$ = annual growth rate of per capita income, and

$\quad$ m $\qquad$ = income elasticity for milk.

Apparent consumption in the base year, 1988, was approximated by adding imports and subtracting exports from the reported milk output of each country in that year.

It was assumed that the population of each country in the terminal year T was given by:

(6) $N_T = N_{1988} (1+n)^{T-1988}$, where

$\quad$ T $\quad$ = 2000 and 2010,

$\quad$ N_T $\quad$ = estimated population in the year T,

$\quad$ N_{1988} = reported population in the base year, 1988, and

$\quad$ n $\qquad$ = estimated growth rate of population.

Combining equations (5) and (6), the total projected milk consumption (TC_T) in the terminal year T is given by:

(7) $TC_T = TC_{1988}\{(1+i*m)(1+n)\}^{T-1988}$.

The income growth rate for each country was estimated using the procedure set out in equations (1) - (3). The population growth rate was then estimated for each country using a similar procedure. These estimated growth rates were substituted in equation (7) to compute the projected consumption levels in the years 2000 and 2010. The projected consumption figures were aggregated to get the regional or subregional figures. In computing the projected consumption using (7), a floor of 0.1 percent and a ceiling of 5 percent were imposed on the annual growth rate of income.

An important assumption underlying the consumption and production projections is that the real price of milk remains unchanged. For production, this assumption is plausible only under the assumption that governments continue to regulate prices as they have in the past, and set prices at levels similar to those prevailing today. Recent policy changes in the developed countries suggest that this assumption is not likely to hold. The supply price elasticity of milk is reasonably high in most countries. For consumption, the assumption of a constant real price is perhaps not so important. Most studies also show that the demand price elasticity of milk is very low. Consequently a change in the relative price of milk will elicit little response in demand.

International Trade

The difference between the projected production and the projected consumption yielded the projected surplus or deficit for each country. These figures were then aggregated to obtain the data for each region.

References

Allen, L.H., Black, A. K., Backstrand, J. R., Pelto, G. H., Ely, R. D., Molina, E., and Chavez, A. 1991. *Food and Nutrition Bulletin* 13 (2):91-105.

Asimov, G. J. and Krouze, N. K. 1937. "The Lactogenic Preparations from the Anterior Pituitary and the Increase in Milk Yield in Cows." *Journal of Dairy Science* 20: 289-306.

Bauman, D. E. 1987. "Bovine Somatotropin: The Cornell Experience." *Proceedings of National Invitational Workshop on Bovine Somatotropin*, St. Louis, Missouri. Pp. 46-56.

Bauman, D. E., Eppard, J. E., DeGeeter, M. J., and Lanza, G. M. 1985. "Responses of High Producing Dairy Cows to Long Term Treatment with Pituitary Somatotropin and Recombined Somatotropin." *Journal Dairy Science* 68: 1352-1362.

Bauman, D. E., Hard, D. L., Crooker, B. A., Partridge, M. S., Garrick, K., Sandles, L. D., Erb, H. N., Franson, S. E., Hartness, G. F., and Hintz, R. L. 1989. "Long Term Evaluation of a Prolonged-release Formulation N-methionyl Bovine Somatotropin in Lactating Dairy Cows." *Journal of Dairy Science* 72: 642-651.

Blake, R. W., Holmann, F. J., Gutierrez, J. and Cevallos, G. F. 1988. "Comparative Profitability of United States Holstein Artificial Insemination Sires in Mexico." *Journal of Dairy Science* 71: 1378-1388.

Blake, R. W., Holmann, F. J., and Stanton, T. L. 1989. "Interacciones entre programas de mejoramiento genético, su rentabilidad y el manejo en México." *Proceedings of Fifth International Conference on Dairy Cattle*, Mexico City.

Boehlje, M., Cole, G. and English, B. 1987. "Economic Impact of Bovine Somatotropin on the U.S. Dairy Industry." *Proceedings of National Invitational Workshop of Bovine Somatotropin*, St. Louis, Missouri. Pp. 167-76.

Brumby, P. 1979. "Cattle Improvement: Imports and Crossbreeding in the Humid Tropics," in D. S. Balaine, ed., *Dairy Cattle Breeding in the Humid Tropics*. Pp. 199-206. Hissar: Haryana Agricultural University.

Buckwell, A. E. 1987. "Structural Impacts of Bovine Somatotropin in the Dairy Sector of the European Community," Working Paper, Department of Agricultural Economics, Wye College, University of London.

Buckwell, A. and Moxley, A. 1990. "Biotechnology and Agriculture." *Food Policy* 15: 44-56.

Burton, J. H., MacLeod, G. K., McBride, B. W., Burton, J. L., Bateman, K., McMillan, I., and Eggert, R. G. 1990. "Overall Efficacy of Chronically Administered Recombinant Bovine Somatotropin to Lactating Dairy Cows." *Journal of Dairy Science* 73: 2157-2167.

Burton, J. H., McBride, B. W., Bateman, K., McLeod, G. K., Eggert, R.K. 1987. "Recombinant Bovine Somatotropin: Effects on Production and Reproduction in Lactating Cows." *Journal of Dairy Science* 70 Supplement: 175.

Butler, L. J. 1995. "rBST and Industrial Dairying in the Sunbelt." Paper presented at 1995 Annual Meeting and Science Innovation Exposition of the American Association for the Advancement of Science, February, 1995.

Butler, L. J. and Carter, H. O. 1988. *Potential Impacts of Bovine Somatotropin on the U.S. Dairy Industry A West Coast Perspective*, Paper No. 88-4, Agricultural Issues Center, University of California.

Buvanendran, V. and P. H. Petersen. 1980. "Genotype-Environment Interaction in Milk Production under Sri Lanka and Danish Conditions." *Acta Agriculture Scandanavica* 30: 369-372.

Chalupa, W. and Galligan, D.T. 1988. "Nutritional Implications of Somatotropin for Lactating Cows." *Journal of Dairy Science* 71 Supplement: 123.

Commission of the European Communities. 1993. "Final Scientific Report of the Committee for Veterinary Medicinal Products on the Application for Marketing Authorization submitted by the Monsant Company for Somatech, 500 mg Sometribove (Recombinant Bovine Somatotropin) for Subcutaneous Injection." Commission of the European Communities, Brussels, January 27.

Cunningham, E. P. 1987. "Crossbreeding - The Greek Temple Model." *Journal of Animal Breeding,* Genetics 104: 2-11.

Cunningham, E. P. 1988. "Biotechnology in Livestock Production," Paper presented at Ninth Agricultural Symposium. Washington, D.C.: World Bank.

Cunningham, E. P. 1989a. "The Genetic Improvement of Cattle in Developing Countries." *Theriogenology* 31: 17-28.

Cunningham, E. P., 1989b, "Formulation of Breeding Plans for Dairy and Dual Purpose Cattle." *Review of Brazilian Genetics* 12 Supplement.

Cunningham, E. P. 1990. "Pilot Trial on the Use of New Reproductive Technology to Improve Milk Production in Developing Countries." Paper prepared for American Breeders Service, DeForest, Wisconsin.

Cunningham, E. P. and Syrstad, O. 1987. *Crossbreeding Bos Indicus and Bos Taurus for Milk Production in the Tropics.* FAO Animal Production and Health Paper 68. Rome: FAO.

Dahl, G. and Hjort, A. 1976. *Having Herds.* Studies in Social Anthropology 2. Department of Social Anthropology, University of Stockholm.

Davidson, B. R. and Martin, B. R. 1965. "The Relationship between Yields on Farms and in Experiments." *Australian Journal of Agricultural Economics* 9: 129-140.

de Haan, C. and Nissen, N. 1985. *Animal Health Services in Sub-Saharan Africa: Alternative Approaches.* World Bank Technical Paper No. 44. Washington, D.C.: World Bank.

Dentine, M. R., and McDaniel, B. T. 1987. "Expected Early Genetic Gain from Selection for Milk Yield in Dairy Cattle." *Theoretical and Applied Genetics* 74: 753-757.

Eddebbarh, A. 1987. "Dairy Development in Morocco: An Overview and Policy Implications," in P. Pinstrup-Andersen, ed., *Economics of Dairy Development in Selected Countries and Policy Implications.* Unpublished manuscript, proceedings of International Food Policy Research Institute conference by the same name, Copenhagen.

Fallert, R., McGuckin, T., Betts, C., and Bruner, B. 1987. *bST and the Dairy Industry: A National, Regional and Farm-level Analysis.* Agricultural Economics Report No. 579. Washington, D.C.: U.S. Department of Agriculture.

Feder, B. 1995. "Wider Use of Cow Drug is Reported." *New York Times,* February 1.

Ferrara, L., Di Liccia, A., Manniti, F., Piva, G. G., Masaero, F. , Fiorentini, L., and Litta, G. 1989. "The Effect of Somidobove (biosynthetic bovine somatotropin) on the production and quality of milk of buffaloes (Bubalus bubalis) raised in Italy." *Asian-Australasian Journal of Animal Sciences* 2: 2-7.

Frankel, J. 1982. *A Review of Bank Financed Dairy Development Projects.* World Bank AGR Technical Note No. 6. Washington, D.C.: World Bank.

Greve, T. 1986. "Practical Aspects of Embryo Transplantation in Cattle." *British Veterinary Journal* 142: 228-232.

Hallberg, M.C., and Cho, W.-J. 1987. "The World Dairy Market: Policies, Trade Patterns and Prospects." A.E. & R.S. Working Paper 191, Department of Agricultural Economics and Rural Sociology, The Pennsylvania State University, August.

Hodges, J. 1985. "Strategies for Dairy Cattle Improvement in Developing Countries," in A.J. Smith, ed., *Milk Production in Developing Countries.* Pp. 198-217. Edinburgh: University of Edinburgh Press.

Hodges, J. 1988. "Genetic Improvement of Livestock in Developing Countries Using the Open Nucleus Breeding System." Paper presented at FAO Regional Workshop on Biotechnology in Animal Production and Health in Asia, Bangkok, Thailand.

Holmann, F. 1990. "Grupos genéticos y sistemas de producción de leche en países tropicales: Experiencias en investigación y programas de desarrollo." Ottawa: International Development Research Centre.

Holmann, F., Blake, R. W., Hahn, M. V., Barker, R., Milligan, R. A., Oltenacu, P. A., and Stanton, T.L. 1990a. "Comparative Profitability of Straightbred and Crossbred Holstein Herds in Venezuela." *Journal of Dairy Science* 73: 2190-2205.

Holmann, F., Blake, R. W., Milligan, R. A., Barker, R., Oltenacu, P. A., and Hahn, M. V. 1990b. "Economic Returns from United States Artificial Insemination Sires in Holstein Herds in Colombia, Mexico, and Venezuela." *Journal of Dairy Science* 73: 2179-2189.

Jarvis, L. S. 1982. "To Beef or Not to Beef: Portfolio Choices of Asian Smallholder Cattle Producers," in J. F. Fine and R. G. Lattimore, eds., *Livestock in Asia: Issues and Policies.* Pp. 29-41. Ottawa: International Development Research Centre.

Jarvis, L. S. 1986. *Livestock Development in Latin America.* Washington, D.C.: World Bank.

Jarvis, L. S. 1991. "Overgrazing and Range Degradation: Is There Scope for Government Control of Animal Numbers?" *East Africa Economic Review* 7: 95-116.

Jasiorowski, H. A., Stolman, M. and Reklewski, Z. 1988. *The International Friesian Strain Comparison Trial: A World Perspective.* Rome: FAO.

Jordan, D.C., Aguilar, A.A., Olson, J.D., Bailey, C., Hartnell, G.F., and Madsen, K.S. 1991. "Effects of Recombinant Methionyl Bovine Somatotropin (Sometribove) in High Producing Cows Milked Three Times Daily." *Journal of Dairy Science* 74: 220-226.

Kaiser, H. M., and Tauer, L.W. 1989. "Impact of Bovine Somatotropin on U.S. Dairy Markets Under Alternative Policy Options." *North Central Journal of Agricultural Economics* 11: 59-73.

Kalter, R. J., Milligan, R., Lesser, W., Magrath, W. and Bauman, D. 1984. *Biotechnology and the Dairy Industry: Production Costs and Commercial Potential of the Bovine Growth Hormone*. A.E. Research Paper 84-22, Department of Agricultural Economics, Cornell University.

Kasonta, J. S. and Nitter, G. 1990. "Efficiency of Nucleus Breeding Schemes in Dual-Purpose Cattle of Tanzania." *Animal Production* 50: 245-251.

Kenney, M., and Buttel, F. 1985. "Biotechnology: Prospects and Dilemmas for Third World Development." *Economic Development and Cultural Change* 16: 61-91.

Klotz, C., Saha, A., and Butler, R.J. 1995. "The Role of Information in Technology Adoption: The Case of rBST in the California Dairy Industry." *Review of Agricultural Economics*, forthcoming.

Kronfeld, D. S. 1993. "Recombinant Bovine Growth Hormone: Its Efficacy, Safety and Impacts on Herd Health Management, Economics, and Public Health," in W. C. Liebhardt, ed., *The Dairy Debate: Consequences of Bovine Growth Hormone and Rotational Grazing Technologies*. Davis, CA: University of California Sustainable Agriculture Research and Education Program.

Larsen, A. and Brown, C. 1987. "The Dairy Policies of the European Economic Community and their Impact on the Dairy Sector of Developing Countries," in P. Pinstrup-Andersen, ed., *Economics of Dairy Development in Selected Countries and Policy Implications*. Unpublished manuscript, proceedings of International Food Policy Research Institute conference by the same name, Copenhagen.

Lemieux, C. M., and Wohlgenant, M. K. 1989. "*Ex Ante* Evaluation of the Economic Impact of Agricultural Biotechnology: The Case of Porcine Somatotropin." *American Journal of Agricultural Economics* 71: 903-914.

Leuning, R., Klemme, R., and Howard, W. 1987. "Wisconsin Farm Enterprise Budgets — Dairy Cows and Replacements, 1987." University of Wisconsin Extension, A 2731, Madison.

Ludri, R.S., Upadhay, R. C., Singh, M., Guneratne, J. R. M., and Basson, R. P. 1989. "Milk Production in Lactating Buffalo Receiving Recombinantly Produced Bovine Somatotropin." *Journal of Dairy Science* 72: 2283-2287.

Madalena, F.E. 1981. "Crossbreeding Strategies for Dairy Cattle in Brazil." *World Animal Report* 38: 23-30.

Madalena, F., Lemos, A. M., Teodoro, R. L., Barbosa, R. T., and Monteiro, J. B. N. 1990. "Dairy Production and Reproduction in Holstein-Friesian and Guzera Crosses." *Journal of Dairy Science* 73: 1872-1886.

Magrath, W. B., and Tauer, L.W. 1985. *The Economic Impact of bGH on the New York State Dairy Sector: Comparative Static Results*. Department of Agricultural Economics Staff Paper No. 85-22. Cornell University, August.

Marion, B. W. and Wills, R. L. 1990. "A Prospective Assessment of the Impacts of Bovine Somatotropin: A Case Study of Wisconsin." *American Journal of Agricultural Economics* 63: 685-99.

Martin, D. L., Knutson, R. G., Blake, R. W., and Tomaszewski, M. A. 1987. "Commercial Feasibility of Embryo Transfer Technology: A Case Study." *Journal of Dairy Science* 70: 2203-2207.

Mbogoh, S. G. 1991. "Dairy Development Schemes in Ethiopia and Kenya," in P. Pinstrup-Andersen, ed., *Economics of Dairy Development in Selected Countries and Policy Implications.* Unpublished manuscript, proceedings of International Food Policy Research Institute conference by the same name, Copenhagen.

McBride, B. W., Burton, J. L., Gibson, J. P., Burton, J. H., and Eggert, R. G. 1990. "Use of Recombinant Bovine Somatotropin for up to Two Consecutive Lactations on Dairy Production Traits." *Journal of Dairy Science* 73: 3248-3257.

McDowell, R. E. 1985. *Meeting Constraints to Intensive Dairying in Tropical Areas.* Cornell International Agriculture Mimeograph 108. Ithaca: Cornell University.

McGuirk, A. M., and Kaiser, H. M. 1991. "bST and Milk: Benefit or Bane?" *Choices,* First Quarter: 20-26.

Minson, D. J. 1981. "Nutritional differences between tropical and temperate pastures," in F. H. W. Morley, ed., *Grazing Animals.* Pp. 143-1157. Amsterdam: Elsevier.

Mukherjee, T. K., ed. 1990. *Proceedings of FAO/UNDP Workshop on Biotechnology in Animal Production and Health in Asia and Latin America.* Rome: FAO.

Nicholas, F. W. and Smith, C. 1983. "Increased Rates of Genetic Change in Dairy Cattle by Embryo Transfer and Splitting." *Animal Production* 36: 341-353.

Nicholson, C.F. 1990. An Optimization Model of Dual Purpose Cattle Production in the Humid Lowlands of Venezuela. M.S. Thesis, Cornell University.

Norman, H. D., and Powell, R. L. 1986. "Pedigree Selection of Dairy Bulls in the United States and Resultant Progeny Tests," *Proceedings of Third World Congress on Genetics Applied to Livestock Production,* Lincoln, Nebraska.

OECD. 1989. *Bio Technology: Economic and Wider Impacts.* Paris: OECD.

Paulino, L. A. 1986. *Food in the Third World: Past Trends and Projection to 2000.* Research Report No. 52, Washington, D. C.: International Food Policy Research Institute.

Pearson, R. E., and Freeman, A. E. 1973. "Effect of Female Culling and Age Distribution of the Dairy on Profitability." *Journal of Dairy Science* 56: 1459-1471.

Perrin, R. K. 1991. "*Ex-Ante* Evaluation of Experimental Technology," Unpublished manuscript, Department of Agricultural and Resource Economics, North Carolina State University, October.

Phipps, R. H., Weller, R. F., Craven, N., and Peel, C. J. 1990. "Use of Prolonged-Release Bovine Somatotropin for Milk Production in British Friesian Dairy Cows." *Journal of Agricultural Science, Cambridge* 115: 95-104.

Phipps, R. H., Madakadze, C., Mutsvangwa, T., Hard, D. L., and Kerchove, G. de. 1991. "Use of Bovine Somatotropin in the Tropics: The Effect of Sometribove on Milk Production of Box Indicus Dairy Crossbred and Bos Taurus Cows in Zimbabwe." *Journal of Agricultural Science, Cambridge* 117: 257-263.

Pinstrup-Andersen, P., ed. 1987. *Economics of Dairy Development inSelected Countries and Policy Implications.* Unpublished manuscript, proceedings of International Food Policy Research Institute conference by the same name, Copenhagen.

Preston, T. R. 1976. "Prospects for the Intensification of Cattle Production in Developing Countries," in A. J. Smith, ed., *Beef Cattle Production in Developing Countries*. Pp. 242-257. Edinburgh: University of   Edinburgh Press.

Preston, T. R. 1977. "A Strategy for Cattle Production in the Tropics."*World Animal Review* 21: 11-17.

Preston, T. R. and Leng, R. A. 1987. *Matching Ruminant ProductionSystems with Available Resources*. Armidale: Penambul Books.

Reed, B. 1994. "For Wages and Benefits, Bigger Dairies May Be Better." *California Agriculture* 48: 9-13.

Roush, W. 1991. "Who Decides About Biotech: The Clash Over Bovine Growth Hormone." *Technology Review* 14: 28-34.

Ruane, J. 1988. "Review of the Use of Embryo Transfer in the Genetic Improvement of Dairy Cattle." *Animal Breeding Abstracts* 56: 437-446.

Saha, A., Love, H. A., and Schwart, R. 1994. "Adoption of Emerging Technologies Under Output Uncertainty." *American Journal of Agricultural Economics* 76: 836-846.

Seidel, Jr., G. E. 1989. "Embryo Transfer and Related Biotechnologies in Cattle." *Beef Cattle Science Handbook*. Pp. 53-58. Bryan, TX: Laney Printing.

Seidel, Jr., G. E. 1986. "Costs and Success Rates with Embryo Transfer," in *Bovine Embryo Transfer: Short Course Proceedings*. Fort Collins: Animal Reproduction Laboratory, Colorado State University.

Seidel, Jr., G. E., and Seidel, S. M. 1981. "The Embryo Transfer Industry." in Brackett, G., Seidel, Jr., G. E., and Seidel, S. M., eds., *New Technologies in Animal Breeding*. pp. 41-80. New York: Academic Press.

Seré, C. 1983. "Primera aproximacíon a una clasificación de sistemas de producción lechera en el trópico sudamericano." *Producción Animal* 8: 110-121.

Seré, C. and Rivas, L. 1987. "Dairy Production in Dual Purpose Herds: Evidence from Tropical Latin America," in P. Pinstrup-Andersen, ed., *Economics of Dairy Development in Selected Countries and Policy Implications*. Unpublished manuscript, proceedings of International Food Policy Research Institute conference by the same name, Copenhagen.

Seré, C., Li Pun, H. H., and Estrada, R.D. 1990. "The Rationale of Developing Dairy Production in the Third World." Paper presented at the 23rd International Dairy Congress, Montreal, Canada.

Shapiro, K., Jesse, E., and Foltz, J. 1991. "Dairy Marketing and Development in Africa." Processed. Department of Agricultural Economics, University of Wisconsin-Madison.

Singh, K. 1987. "Dairy Development In India, 1970-1985: An Overview, Experiences, and Lessons," in P. Pinstrup-Andersen, ed., *Economics of Dairy Development in Selected Countries and Policy Implications*. Unpublished manuscript, proceedings of International Food Policy Research Institute conference by the same name, Copenhagen.

Slade, R. 1987. "The Impact of Operation Flood: A Review of Socioeconomic Evidence," in P. Pinstrup-Andersen, ed., *Economics of Dairy Development in Selected Countries and Policy Implications*. Unpublished manuscript, proceedings of International Food Policy Research Institute conference by the same name,

Copenhagen.

Smith, A. J., ed. 1985. *Milk Production in Developing Countries.* Edinburgh: Center for Tropical Veterinary Medicine.

Smith, C. 1988a. "Applications of Embryo Transfer in Animal Breeding." *Theriogenology* 29: 203-212.

Smith, C. 1988b. "Genetic Improvement of Livestock in Developing Countries Using Nucleus Breeding Units." *World Animal Review* 65: 2-10.

Stennes, B. K. 1989. Bovine Somatotropin and the Canadian Dairy Industry: An Economic Analysis. M.S. Thesis, University of British Columbia.

Stranden, I., Maki-Tanila, A., and Mantysaari, E. A. 1991. "Genetic Progress and Rate of Inbreeding in a Closed Adult MOET Nucleus Under Different Mating Strategies and Heritabilities." *Animal Breeding Genetics* 108: 410-411.

Tuan, F. C. 1987. *China's Livestock Sector.* FAS Economic Report No. 226, Economic Research Service, United States Department of Agriculture.

Turton, J. D. 1981. "Crossbreeding of Dairy Cattle — A SelectiveReview." *Animal Breeding Abstracts,* 49: 293-300.

Turton, J. D. 1985. "Progress in the Development and Exploitation of New Breeds of Dairy Cattle in the Tropics," in A.J. Smith, ed., *Milk Production in Developing Countries.* Pp. 218-239. Edinburgh: University of Edinburgh Press.

Vaccaro, L. de. 1979. "The Performance of Dairy Cattle Breeds in Tropical Latin America and Programmes for Their Improvement." in D.S. Balaine, ed., *Dairy Cattle Breeding in the Humid Tropics.* Hissar: Haryana Agricultural University.

Vaccaro, L. de. 1994. "Technology for Development in the Area of Animal Breeding," in J. R. Anderson, ed., *Agricultural Technology: Policy Issues for the International Community.* Pp. 435-450. Wallingford: CAB International.

Vaccaro, R. 1988. "Metodología apropiada en la evaluación del valor genético de animales en poblaciones cruzadas de doble propósito." *Ciencia y Tecnología de Venezuela* 5: 73-82.

Van Soest, P. J. 1982. *Nutritional Ecology of the Ruminant.* Ithaca, NY: Comstock Publishing.

Vercoe, J.E. and J. E. Frisch. 1980. "Animal Breeding and Genetics with Particular Reference to Beef Cattle in the Tropics," in L.S. Verde and A. Fernandez, eds., *Proceedings of the IV World Conference on Animal Production.* Pp. 452-464. Buenos Aires: Argentine Association of Animal Production.

Walshe, M. J., Grindle, J., Nell, A., and Bachmann, M. 1991. *Dairy Development in Sub-Saharan Africa.* World Bank Technical Paper No. 135. Washington, D.C.: World Bank.

Wideman, D. 1982. "The Genetic Basis for Embryo Transfer," in Donaldson, L. E., ed., *Embryo Transfer in Cattle.* Rio Vista International, Inc.

Wilkins, J., Ali, J.A., and Vaca Díaz, C. 1979. "El cruzamiento para la producción de leche en los llanos bolivianos." Paper presented at VII Reunion, Asociación Latinoamericana de Producción Animal. Panama.

Winrock International. 1992. *Assessment of Animal Agriculture in Sub-Saharan Africa* Arlington: Winrock International.

Zepeda, L. 1990. "Predicting Bovine Somatotropin use by California Dairy Farmers." *Western Journal of Agricultural Economics* 15: 55-62.

About the Book and Author

Biotechnology is expected, by many observers, to have a significant impact on the world dairy industry over the next decade. In this timely volume, Lovell Jarvis analyzes the potential effect of two biotechnologies—multiple ovulation and embryo transfers (MOET) and recombinant bovine somatotropin (rbST—on the dairy industry around the world.

According to Jarvis's research, the effects of these two technologies will vary greatly between the developed and developing nations. He predicts that the technologies will be most profitable for the developed nations, where their use will increase milk production and strengthen their positions in dairy export markets. Developing country dairy sectors will probably lose from the use of these two biotechnologies, as their own international trade position will be weakened, though their own consumers should benefit.

Jarvis concludes his study with a look at alternative approaches that might improve the competitive position of developing countries in the dairy sector.

Lovell S. Jarvis is professor of agricultural economics at University of California at Davis.

Index